BELOW FREEZING
Elegy for the Melting Planet

Donald Anderson
FOREWORD BY Aritha van Herk

University of New Mexico Press | Albuquerque

Library of Congress Cataloging-in-Publication Data
Names: Anderson, Donald, 1946 July 9– author.
Title: Below freezing: elegy for the melting planet / Donald Anderson;
foreword by Aritha van Herk.
Description: Albuquerque: University of New Mexico Press, 2018. |
Includes bibliographical references.
Identifiers: LCCN 2017060738 (print) | LCCN 2018002971 (e-book) |
ISBN 9780826359841 (e-book) | ISBN 9780826359834 (pbk.: alk. paper)
Subjects: LCSH: Climatic changes—Popular works.
Classification: LCC QC903 (e-book) | LCC QC903 .A5354 2018 (print) |
DDC 363.738/74—dc23
LC record available at https://lccn.loc.gov/2017060738

Cover photograph by John Salvino on Unsplash
Designed by Felicia Cedillos
Composed in Minion Pro 10.25/14.25
Maps and illustrations © Benjamin Busch. Used by permission.

For the grandchildren, yours and mine.

Early morning, sun just up, late autumn.
A sudden snow surprises the sparrows and willows.
—ROO BORSON, FROM "CEDARVALE DIARY"

Foreword

Aritha van Herk

No snowflake in an avalanche ever feels responsible.

—VOLTAIRE

In the heat of summer, I recall the deep cold of winter, its chill invitation, its dark embrace, its brisk seductions. I relish the snow season, wintertide's tone. It is the season of hibernal intensity, of snowflake words in an avalanche of stories.

I live on the lip of the foothills, the submontane transition zone rising toward the Rocky Mountains. My climatic area is where the "Alberta Clipper" originates, the much-reviled winter storm that develops on the east side of the Canadian Rockies and then sweeps into a long, mean, low-pressure system, sailing south to catch the jet stream and blasting central and eastern North America with cold and snow. Keen winds and plummeting temperatures assign signature to that weather, although ironically, Alberta will often bask in a contrastingly warm chinook. There resides the beautiful irony of cold: it surprises those who assume they understand its complexity.

We humans think we know the nature of freezing. We see it as immobility, or a metaphor, fright or confinement. We fail to confront, in our protected environments of heated houses and insulated life, the intimate nature of cold, that it can invade and inveigle, that it can cajole and kill. Our own relationship to snow and cold, to freezing and the difficulties it proposes, has become distant, as remote as antiquity. We fail to remember the corporeality of exposure. We take for granted the contemporary benefits and usages of cold: air conditioning, flash freezing, refrigerated transportation, vending machine ice for tailgate parties. We believe we have subzero under control. And humans blow hot and cold, complain about the weather, then expend energy worrying about climate change and the ice melting.

This book, this collage-ledger, *Below Freezing*, is a rich, thoughtful, and beautiful compendium of readings, interpretations, digressions, anecdotes, facts, and citations celebrating cold's complexity. It gathers together poetry and myth, science and art, history and futurity. It links Shakespearean and biblical and literary references in an assemblage that enriches and intensifies the usual stories of frost to create an entirely new work, texts and intertexts suggesting by their juxtaposition a complex and fresh unfolding of the significance and eloquence of cold. *Below Freezing* invites readers to explore and encounter different versions, narrative and poetic, of ice and snow, cold and survival. The multiple quotes and allusions, set together with Donald Anderson's own writing, effect a chiaroscuro of texture and history, warning and depiction. The result achieves an intricate and moving exploration combining scientific and literary exploration and suggestion. It enthralls like a detective story, a meditation, and an exhortation. To read these fragments together, in this way, is quite simply breathtaking.

Cold is first of all experience, a sensation acquired and intensified by an absence of sensation, the benumbing of freezing itself anesthetizing. The feeling of frostbite, the fierce dehydration of intense cold, the brisk absence of breath that cold can provoke, taken together depict an element much more complex than incipient danger or death. Hypothermia, or the necrotization of exposed flesh and the potential perishing of life from cold, evokes antique times, but blizzards and avalanches, extreme windchill, and the bitter purity of cold can still remind humans that we might think we manage freezing, but more likely freezing manages us. Exposed flesh is a human fallibility that cannot be escaped. In fact, cold weather is responsible for death far more than other extreme weather events, like floods and hurricanes.

But while the threat of desperation and death is present, what is most compelling is the personal framing here, how we might transform winter from a dark and isolate destination. Death is certainly a trope associated with cold, but Anderson draws attention to Lueders's beautiful phrase, "the deliberation of winter," as an alternative to "the dead of winter." In this engaging assemblage, the "cosmic insignificance" of humans is not an unhappy reminder, but a placement of order and balance.

The language of cold snaps and echoes throughout this work, like a jazz improvisation. Cold fronts and snow days, windchill factors and blizzards and ice rime evoke a poetry of consequence, language working with concept to achieve connection. The soundtrack of snow and the soundtrack of cold perform an orchestrated composition here, texture and pattern working with meaning to achieve relativity, sensory incentive. Because freezing and weather perform this intricate syncopation, the performers in this complex score are often meteorologists. They speak a language of their own domain, and the people who hear warnings do not always understand how to interpret them. *Below Freezing* unfolds this precise diction, opens up the meaning of windchill and how it is measured, the reverberation of Siberian Highs and polar vortexes. Stories of killer storms and their legendary aftermaths play a compelling role. And beneath is a powerful recognition that the weather going on within us may be more important than the weather without, except that we do not encounter it as purely as we used to. Included too is the blizzard of history, how fighting the weather is more difficult than combat with any enemy. Compellingly, the weather resists logic but wields enormous power, and readily defeated both Napoleon's troops and the German army at Stalingrad, while cold was Stalin's method of punishment in the Gulag too, for it is a prison no one can escape from.

Just a short time ago, a one-trillion-tonne iceberg, the size of Delaware, or Prince Edward Island, broke away from Antarctica. Climate change and global warming discussions attend the event, measures of our increasing awareness of human cause and effect. Antarctica, a continent epitomizing all that ice and cold can be, is freezing's last stronghold, lying beneath a solid shelf of ice, in places three miles thick. The Arctic sea ice and its lineaments play a role here too, more accessible than Antarctica but still mysterious, still a measure of melt and extent. Ice can grow and shrink, it can float and it can capsize, it can be a bridge or a barrier, it is a living, moving book.

And ice has shaped the contours of North America. The most recent (50,000 to 100,000 years ago) was the Wisconsin Glacial Episode, radically changing the geography north of the Ohio River. The Labrador ice sheet shaped the position of the Great Lakes. The rich soil of the Midwest is mostly glacial in origin. Glaciers and glacial outwash rivers are the source of water for millions of people. We are the children of ice ages, and if we imagine that we can escape that etching, we spurn our mnemonic shaping. Ice-age life persists, and even plants, a flower from an ice-age seed, have been grown.

The pull of the polar has been the source of adventures and magnified adventure stories, from the failed Franklin expedition to the summiting of Everest. Both explorers, like Peary and Shackleton and Scott and Amundsen, and daydreamers, who imagine the joy of discovery, fail to grapple with survival's intense demands, how people persist without sufficient caloric intake, the extremes of darkness and light, and, on Everest, how depleted oxygen, which impairs thinking and physical action, also diminishes the body's ability to manage cold. Still, the allure persists, that enticement of the unknown, the formidable, the arduous. From explorers and adventurers to ordinary people who try to comprehend this mysterious world, polarity captures our desire.

Medicine has used cold to its advantage, and cryotherapy, cryopharmaceutical experiment, icing, and transplantation have all benefitted from the slowed metabolism, the strange purity proposed by cold. Humans have adapted to its potential for preservative and sport, the powerful notion of liquid turned solid, and the crystalline shapes of water vapor's destination, snow. Is cold a desolate factor, or does it propose its own life, its own thermometer of what matters? It essentializes survival, meeting human numbness and oblivion with a promise and a challenge. Lying down to sleep in a blizzard is a fine way to die, and there is no dazzle more brilliant than the sky in a polar night.

Below Freezing offers a new approach to some familiar ideas and an unusual merging and juxtaposition of this mesmerizing subject. Reading this "compilation" is as compelling as any adventure tale. Readers interested in climate change will read this book to find answers; discerning readers will happily undertake this voyage of discovery to the physical configuration of the earth we occupy. *Below Freezing* is a story for a literary audience, an environmental audience, a scientific audience, and an audience interested in tales of extremity and exploration, adventure and conflict. The way this chronicle unfolds and the combination of texts with the author's own writing is captivating. Brilliantly, he describes the moment when, like all children, he was unable to resist the challenge of tasting metal in the cold: "the mercury registered 30 below when I chaperoned my experiment." Unconventionally, *Below Freezing* is not a book of facts, although there is material that is interested in "facticity." It is a dance with possibility and history, the two together performing an elegant bricolage.

As a Canadian, I hold this meditation and compendium close. After all, I live in a polar metaphor, like the words of Gilles Vigneault's song, "Mon pays." "My country is not a country, it is winter," "it is a land of snowstorms," he says. But all denizens of earth, those who know desert and lush rain forest and continental and tropical climes, will find in this collection a mesmerizing tale, an isomorphic sculpture impossible to forget. *Below Freezing* is not merely hypothermic; it is hypnotic in its colours and textures, powerfully incremental. This riveting work pulls the reader along with the writer's own exploration of ice and its multiple characteristics. All readers will be entranced, without fail, for ice and cold exert a power as profound as any Ancient Mariner.

Calgary, Alberta
Canada

PART I

God fashioned the ship of the world carefully.
With the infinite skill of an all-master
Made He the hull and the sails,
Held He the rudder
Ready for adjustment.
Erect stood He, scanning His work proudly.
Then—at fateful time—a wrong called,
And God turned, heeding.
Lo, the ship, at this opportunity, slipped slyly,
Making cunning noiseless travel down the ways.
So, that forever rudderless, it went upon the seas
Going ridiculous voyages,
Making quaint progress,
Turning as with serious purpose
Before stupid winds.
And there were many in the sky
Who laughed at this thing.

—STEPHEN CRANE,
from *The Black Riders and Other Lines*

*At its low point during the summer, the Arctic sea ice is on average
820,000 square miles smaller than it was 18 years ago [since the Kyoto
Protocol], according to the National Snow and Ice Data Center. That's
a loss equal in area to Texas, California, Montana, New Mexico and
Arizona combined. . . . The West Antarctica and Greenland ice sheets
have lost 5.5 trillion tons of ice, according to Andrew Shepherd of
Leeds, who used NASA and European satellite data.*

—THE DENVER POST, November 30, 2015

Prologue

That mittens outwork gloves was a lesson learned, growing up in Montana—
sure knowledge more aptly confirmed when an Air Force assignment landed
me at a radar site in Alaska, three degrees north of the Arctic Circle. There we
were issued mittens top-covered with animal hair—caribou, I believe. That
way you could swipe at your streaming nose in frigid cold to be able, in seconds, to remove the collected frozen matter by striking the fur with your other
paw. Not encountering scientific explanations as to why mittens perform in
their superior way, I have had to fall back on my dead father's stated notion
that mittens worked to warm fingers in the way that Eskimos sleep together
naked beneath seal hides. "And by the way," he went on, "should you visit, the
husband will invite you to sleep with his undressed wife to keep you warm in
the igloo." It was a friendly custom Eskimos maintained, he claimed. As a kid,
it didn't occur to me to ask where he had received this cold weather info. As
an adult in Alaska, though, it became expressly clear that we are creatures of
restricted temperature and that flesh of the human sort is in particular sus-
ceptible, dependent upon exposure, to virtually any mark below 32°F.

In Fairbanks, Alaska, the weathermen wait until the temperature registers
-50°F before issuing "extreme cold" warnings. In Miami, they wait until the
temperature falls to +50°F before announcing the opening of shelters due to
cold.

When Air Florida Flight 90 crashed on its January 13, 1982, takeoff attempt at Washington National Airport, the bridge it struck before plunging into the Potomac was less than two miles from the White House and within view from both the Jefferson Memorial and the Pentagon. The airport had just reopened following a heavy snowstorm, and the bridge was jammed with traffic: government workers being released due to weather. Flight 90 had been airborne for thirty seconds when it hit six cars and a truck on the bridge, tearing away ninety-seven feet of bridge rail and forty-one feet of bridge wall before its dive into the iced-over river. In total, seventy-eight people died, including four on the bridge. Five passengers from the plane survived. Of the seventy-four Air Florida passenger fatalities, only one drowned—the survivor who insistently gave up rescue lines to other survivors finally succumbed to hypothermia and perished. The bridge was renamed the "Arland D. Williams Jr. Memorial Bridge" in his honor.

The blizzard of January 12, 1888, known as "the Schoolchildren's Blizzard" because so many of the victims were children caught out on their way home from school, became a marker in the lives of the settlers, the watershed event that separated before and after. The number of deaths—estimated between 250 and 500—was small compared to that of the Johnstown Flood that wiped out an entire industrial town in western Pennsylvania the following year or the Galveston hurricane of 1900 that left more than eight thousand dead. But it was traumatic enough that it left an indelible bruise on the consciousness of the region. The pioneers were by and large a taciturn lot, reserved and sober Germans and Scandinavians who rarely put their thoughts or feelings down on paper, and when they did avoided hyperbole at all costs. Yet their accounts of the blizzard of 1888 are shot through with amazement, awe, disbelief. There are thousands of these eyewitness accounts of the storm. Even those who never wrote another word about themselves put down on paper everything they could remember about the great blizzard of 1888. Indeed, it was the storm that has preserved these lives from oblivion. The blizzard literally froze a single day in time. It sent a clean, fine blade through the history of the prairie. It forced people to stop and look at their existences—the earth and sky they had staked their future on, the climate and environment they had brought their children to, the peculiar forces of nature and nature's God that determined whether they would live or die.

—DAVID LASKIN, *The Children's Blizzard*

4

The National Transportation Safety Board determined that the Air Florida crash was the consequence of pilot error. The captain and his first mate were—despite the weather—essentially responsible for seventy-eight deaths and the rippling effect of such loss. For starters, the pilots failed to engage the engine's internal ice protection systems, used reverse thrust in a snowstorm prior to takeoff (sucking in large amounts of storm debris), and failed to abort takeoff after detecting power problems during taxiing and visually identifying ice and snow buildup on the wings. Earlier, contrary to flight manual guidance, the pilots had attempted to "de-ice" the aircraft by positioning their plane near the exhaust of the aircraft ahead in line. This violation may have only pushed melted ice to the trailing porting of the wings to refreeze. Though both pilots had thousands of hours of flight time, neither was experienced in flying in snowy and cold weather. The captain had made only eight takeoffs and landings in snowy conditions. The copilot had flown in snow only twice.

On January 12, 1888, a blizzard broke over the center of the North American continent. Out of nowhere, a soot gray cloud appeared over the northwestern horizon. The air grew still for a long, eerie measure, then the sky began to roar and a wall of ice dust blasted the prairie. Every crevice, every gap and orifice instantly filled with shattered crystals, blinding, smothering, suffocating, burying anything exposed to the wind. The cold front raced down the undefended grasslands like a crack unstoppable army. Montana fell before dawn; North Dakota, during morning recess; Nebraska as school clocks rounded toward dismissal. In three minutes the front subtracted 18 degrees from the air's temperature. Then evening gathered in and temperatures kept dropping steadily, hour after hour, in the northwest gale. Before midnight, windchills were down to 40 below zero. That's when the killing happened. By morning on Friday the thirteenth, hundreds of people lay dead on the Dakota and Nebraska prairie, many of them children who had fled—or been dismissed from—country schools at the moment when the wind shifted and the sky exploded.

—Ibid.

The T. E. D. Schusters' only child was a seven-year-old boy named Cleo who rode his Shetland pony to the Westpoint school that day and had not shown up on the doorstep by two p.m., when Mr. Schuster went down into the root cellar, dumped purple sugar beets onto the earthen floor, and upended the bushel basket over his head as he slung himself against the onslaught in his second try for Westpoint. Hours later Mrs. Schuster was tapping powdered salt onto the night candles in order to preserve the wax when the door abruptly blew open and Mr. Schuster stood there without Cleo and utterly white and petrified with cold. She warmed him up with okra soup and tenderly wrapped his frozen feet in strips of gauze that she'd dipped in kerosene, and they were sitting on milking stools by a red-hot stove, their ankles just touching, only the usual sentiments being expressed when they heard a clopping on the wooden stoop and looked out to see the dark Shetland pony turned gray and shaggy-bearded with ice, his legs as wobbly as if he'd just been born. Jammed under the saddle skirt was a damp, rolled-up note from the Scottish schoolteacher that said, Cleo is safe. The Schusters invited the pony into the house and bewildered him with praises as Cleo's mother scraped ice from the pony's shag with her own ivory comb, and Cleo's father gave him sugar from the Dresden bowl as steam rose up from the pony's back.

—RON HANSEN, "Wickedness"

Television weathercasters like to say that windchill is what the weather feels like. Using the 2001 windchill index, when the wind is blowing 30 miles an hour at a temperature of 25 degrees, it feels like 8 degrees. "Feels like" is a fuzzy term for an exact transaction. What windchill means is that it's irrelevant that the thermometer reads 25 degrees: If the wind is blowing at 30 miles an hour, the exposed parts of your body are losing heat at the rate that they would if the temperature were in fact 8 degrees.

When the Schweizer boys left school late in the morning, the windchill was about 5 degrees above zero. At 9 P.M., four hours after the sun set, the windchill had dropped to 40 below zero. In conditions like that, exposed flesh freezes in ten minutes.

Ten minutes to turn warm skin and blood to ice. The five boys had been outdoors by that point for over nine hours.

—DAVID LASKIN, *The Children's Blizzard*

Hypothermia is a medical emergency that occurs when your body loses heat faster than it can produce it. Normal body temperature is around 98.6°F. Hypothermia occurs as your body temperature passes below 95°F. When your body temperature drops, your heart, nervous system and other organs stop working effectively. Untreated, hypothermia can eventually lead to failure of the heart and respiratory system. Hypothermia is most often caused by exposure to cold weather or immersion in a cold body of water.

For the survivors of Flight 90, the below-freezing waters of the Potomac and its heavy ice made swimming all but impossible. When Arland Williams—the potential sixth and final survivor—finally disappeared beneath the icy surface, he had been in the water for twenty-nine minutes.

Even at six o'clock that evening, there was no heat in Mathias Aachen's house, and the seven Aachen children were in whatever stockings and clothing they owned as they put their hands on a Hay-burner stove that was no warmer than soap. When a jar of apricots burst open that night and the iced orange syrup did not ooze out, Aachen's wife told the children, You ought now to get under your covers. While the seven were crying and crowding onto their dirty floor mattresses, she rang the green tent cloth along the iron wire dividing the house and slid underneath horse blankets in Mathais Aachen's gray wool trousers and her own gray dress and a ghastly muskrat coat that in hot weather gave birth to insects.

Aachen said, Every one of us will be dying of cold before morning. Freezing here. In Nebraska.

His wife just lay there, saying nothing.

Aachen later said he sat up bodingly until shortly after one p.m., when the house temperature was so exceedingly cold that a gray suede of ice was on the teapot and his pretty girls were whimpering in their sleep. You are not meant to stay here, Aachen thought, and tilted hot candle wax into his right ear and then his left, until he could only hear his body drumming blood. And then Aachen got his Navy Colt and kissed his wife and killed her. And then walked under the green tent cloth and killed his seven children, stopping twice to capture a scuttling boy and stopping once more to reload.

—RON HANSEN, "Wickedness"

It's been widely reported that the "Russian Winter" was primary in bullying Napoleon's *Grand Armée* retreat from Moscow in 1812. Also reported is that a French surgeon erroneously instructed the frostbitten to rub snow on their affected parts, a procedure that proved nearly as damaging as direct heat from fire.

A five-thousand-year-old pre-Columbian mummy discovered in the Chilean mountains offers the earliest documented evidence of frostbite.

Modern medicine calls for painkillers to dull the rewarming of frostbitten flesh in a whirlpool bath of 104–107°F, hydration with warm fluids, a high-protein diet, and a course of antibiotics. Besides rubbing frostbite with snow (a slight improvement over the scorching of frozen flesh over roaring campfires), Napoleon's troops worsened their damaged feet and hands, ears, noses, and genitals by thawing extremities, only to have them refreeze.

In 1925, a serum run to Nome involving 20 mushers and some 150 sled dogs relayed diphtheria antitoxin 674 miles across the US territory of Alaska in five and a half days, a record which has never been broken, saving the small city of Nome and the surrounding native communities from the epidemic. Before viable arctic aircraft in the 1930s and the snowmobile in the 1960s, the sled dog was the primary transportation and means of communication in sub-arctic areas around the world.

The resurgence of recreational mushing is the result of the popularity of the Iditarod Trail Sled Dog Race, which honors the history of dog mushing and commemorates the serum run. During the original Nome run, temperatures dropped to more than -60°F. At times, the winds pushed the windchill to -85°F.

Entr'acte

The radar base camp where I was stationed in central Alaska sat beside the Indian River that connected to the Koyukuk, a primary northern tributary to the mighty Yukon. There was an Indian fishing village, where the Koyukuk met the Yukon. During the winter, when the Indian River froze, the fishermen would drive their snow machines up the frozen river to play cards and drink liquor. They'd drink up then head back to their village, swerving and whooping in the refrigerated dark. There was an older Indian, who when he arrived, came by way of dogs and sled. He'd drink, then go outside to sleep with his animals. It could be 40 or 50 below and he'd trudge out. You'd hear the dogs yipping in the mornings—at four or five o'clock—as he tossed them frozen fish. Once when he'd mushed up for a night, I asked, "Why don't you drive a snowmobile?" He looked at me. Then: "If your snowmobile dies, what are you going to do—eat the engine?"

In point of fact, Napoleon's army suffered as much damage from the heat of the Russian summer as from the rigors of the winter. Tens of thousands of cavalry and artillery horses died before Napoleon ever reached Moscow; tens of thousands of men dropped out of the ranks through sickness and heat exhaustion before the Battle of Borodino was fought. The hot weather of July and August was as much to blame for Napoleon's defeat as the frosts of November and December. Indeed, the conditions pertaining at the outset of the retreat were far more favorable than might have been hoped for. The first severe frosts were encountered only on November 12, and these gave way to an unseasonable thaw that proved even more embarrassing to the French as they approached the crucial Berezina crossing, producing muddy roads, impassable countryside and swollen waterways instead of firm going and frozen rivers. . . . Indeed, the real effects of winter were experienced only after December 4, when the temperature plunged many degrees below zero. Well before that date, the *Grande Armée* had been reduced to a shadow of its former self. However, it is true that the cold served to increase the scale of the disaster during the final state of the retreat when the strategic outcome had already been long decided.
—DAVID G. CHANDLER, *The Campaigns of Napoleon*

"Of what the world is really made,
we are such easy prey," said Nietzsche.
Black ice, reckless driving, loose dogs

weaving through rigged parking meters
lining the curb like a kickbacked jury.
I kept the bladed key coldly cranking
my morgue-bound German standard-shift,
hoofed my frozen boot at the grudging
pedal, bred the engine down to clicks,
a Gatling gun jammed in a nuclear age.
No spark, fuel, the block bled out black;
pressed to all fours, I nuzzled against the fender
to make sure, crouched above the motor
silent as ice . . . as if there were a heart to gnaw.

—M. K. SUKACH, "Stalled"

The world record for the most snow in one year is accorded to Mount Baker in Washington State. The Mount Baker Ski Area reported 1,140 inches (95 feet of snowfall for the 1998–1999 season). Mount Baker Ski Area holds the record for one month of snow, too: 304 inches.

Of the 655,000 soldiers who marched toward Russia in 1812, at least 570,000 were lost to the battlefield or to illness and exposure. Napoleon was unable to mount such an army again. The falsehood that the French had been defeated by the "Russian Winter" was initiated by the Emperor himself. "My army has had some losses," he modestly admitted to the Senate on December 20, "but this was due to the premature rigors of the season." Napoleon had abandoned his troops in the field on December 5 to return to Paris. He accomplished that trip in a horse-drawn coach.

—DAVID G. CHANDLER, *The Campaigns of Napoleon*

Though little is known about the blizzard that struck Iran in February of 1972, its reported death toll of four thousand ranks it as one of the deadliest snowfalls of all time. Whole communities were wiped out. For instance, near the border with Turkey, the village of Sheklab with its one hundred inhabitants was buried in twenty-five feet of snow.

In 1938, in my hometown of Butte, Montana, it snowed nonstop for eighty hours.

Before the fateful Air Florida Flight 90 crash, Washington National Airport had been closed by a heavy snowstorm. After leaving the gate, the aircraft waited in a taxi line for 49 minutes before reaching the takeoff runway. All the while snow and ice accumulated on the wings that had been improperly de-iced. Heavy snow was falling during the take-off roll. The last words recorded:

First officer: "We're going down, Larry . . ."
Captain: "I know it."

Earlier in the recorded conversation:

First officer: "It's a losing battle trying to de-ice these things. It gives you a false feeling of security, that's all it does."
Captain: "Well, it satisfies the Feds."

Turns out it was against Federal regulation to sell unopened liquor to the visiting Indians at the radar site in Alaska. We could, though, sell them all they could drink during their visit before departing into the arctic dark and cold.

> The snow comes down hard.
> There's no road, just white,
> and no other car lights
> beyond the reflection of my hood.
> Turn-offs are dark to the right
> or left. For miles, no red
> taillights, just the on beat
> off beat of wipers
> driving me to sleep in a bed
> of black ice, rocking my gentle skid.
> Two good snow tires aren't enough.
> I think chains could hold
> the road if I had them.
> The steering swerves, no friction
> like smooth moments before a dream.

The car cradles my senses numb, I wish
for soft snow banks to lend a shoulder.
I've forgotten the sounds of everything
reviewed in front of me;
my life snowing, no turn-off now,
no time to stop on ice.

—WILLIAM HOCHMAN, "Drive From Butte"

Three Heavyweight Boxing Champions: John L. Sullivan, the Man himself, as well as Jim Jeffries and Bob Fitzsimmons all fought bouts in Butte. Holding the record for the lowest winter temperature (-61°F) in the contiguous United States, Butte is where J. Edgar Hoover assigned FBI agents who particularly nettled him.

Let me hear the clatter of hailstones on icebergs, and not the dull
tramp of these plodders, plodding their dull way from their cradles to
their graves.

—HERMAN MELVILLE, *White Jacket*

I loved hearing my grandparents tell stories about their childhood in Spain. I especially loved hearing about snow and ice on Christmas Eve. I'd ask them hundreds of questions: What does snow look like? What does it feel like when you touch it? Does snow smell like the frost in our freezer? What does it feel like to wear coats and hats all the time? Did you ever make a snowman? Did you ever have snowball fights?

I thought our Christmases in Cuba were inferior because we didn't have snow. Christmas was all about snow, and here we were, eating Christmas Eve dinner in our shirtsleeves, with palm trees waving in the wind outside. We Cubans were getting cheated out of the real Christmas.

—CARLOS EIRE, *Waiting for Snow in Havana*

And Franky Gorky was up like a dart on Christmas morning, waiting at the window for his Grandma Gorky, and when she came he did the first bad thing of his life, he ran out of the house without permission and headed across the street where Grandma Gorky had parked because Christmas visitors all over the neighborhood had taken all the parking places on the Gorkys' side, slipped on the ice, and got rubbed out by a drunk driver.

That's what happens, said my father, when people take other people's parking places.

That's what happens, said my mother, when you don't look both ways.

What happens is, if you're the nicest kid in the whole universe, then you have to die.

This is what happens when you try to explain something.

—CHUCK ROSENTHAL, "The Nicest Kid in the Universe"

Entr'acte

As that young officer at the radar site in Alaska, I refused to exchange an unopened bottle of Jim Beam with the old Indian for a pair of seal mukluks. How now not to feel that I should have taken the offer, Federal regs aside?

> My favorite story that night was the one my grandfather told about finding a wolf who had frozen to death. "There he was, the beast, totally stiff, hard as a rock. I tried to bend his legs, but they were like steel. So I picked him up and threw him, and he sounded just like a rock when he hit the ground. And one of his ears broke off. Just snapped right off, cracked as easily as a mirror, it did." I tried to imagine air cold enough to freeze a wolf solid, but couldn't.
>
> —CARLOS EIRE, *Waiting for Snow in Havana*

He was a newcomer in the land, a *chechaquo*, and this was his first winter. The trouble with him was that he was without imagination. He was quick and alert in the things of life, but only in the things, and not in the significances. Fifty degrees below zero meant eighty-odd degrees of

13

frost. Such facts impressed him as being cold and uncomfortable, and that was all. It did not lead him to meditate upon his frailty as a creature of temperature, and upon man's frailty in general, able only to live within certain narrow limits of heat and cold; and from there on it did not lead him to the conjectural field of immortality and man's place in the universe. Fifty degrees below zero stood for a bite of frost that hurt and that must be guarded against by the use of mittens, ear-flaps, warm moccasins, and thick socks. Fifty degrees below zero was to him just precisely fifty degrees below zero. That there should be anything more to it than that was a thought that never entered his head. . . .

At the man's heels trotted a dog, a big native husky, the proper wolf-dog, gray-coated and without any visible or temperamental difference from its brother, the wild wolf. The animal was depressed by the tremendous cold. It knew that it was no time for travelling. Its instinct told it a truer tale than was told to the man by the man's judgment. In reality, it was not merely colder than fifty below zero; it was colder than sixty below, than seventy below. It was seventy-five below zero. Since the freezing-point is thirty-two above zero, it meant that one hundred and seven degrees of frost obtained. The dog did not know anything about thermometers. Possibly in its brain there was no sharp consciousness of a condition of very cold such as was in the man's brain. But the brute had its instinct. It experienced a vague but menacing apprehension that subdued it and made it slink along at the man's heels, and that made it question eagerly every unwanted movement of the man as if expecting him to go into camp or to seek shelter somewhere and build a fire. The dog had learned fire, and it wanted fire. . . .

—JACK LONDON, "To Build a Fire"

I would look at maps of the world and long for northern latitudes. I actually used to think that the farther north you went on the globe, the purer things became. I remember stretching out on the cold, white marble floor of my house, the closest I could get to ice, and staring at maps for hours and hours, wondering what it would be like anywhere north of Cuba, and especially above latitude forty-five degrees North. Or even better, fifty degrees North, no eighty degrees North. How I wanted to live in Norway, Sweden, Finland, Iceland, Greenland,

Alaska, Siberia, Yukon, Baffin Island. The North Pole. All that white ice, all that snow and cold air. So pure, so good. Snow was grace itself, falling from heaven; it didn't hide evil, but vanquished it. And I longed for it, fervently, there in Havana.

—CARLOS EIRE, *Waiting for Snow in Havana*

"You are stronger than one dog," Tallak told us. "One dog, you can pin him down, you can make him do what you want. Two dogs, maybe. Four dogs can pull a truck from a ditch. The sled's brake is a joke. It is a suggestion. When the dogs are together, you have no chance of controlling them, unless they choose to please you."

He was right. It wouldn't be long before I saw a six-dog team uproot a birch tree that they were tied to and take off down the trail, tree and roots bouncing behind them.

—BLAIR BRAVERMAN, *Welcome to the Goddamn Ice Cube*

Yes, the newspapers were right: snow was general all over Ireland. It was falling on every part of the dark central plain, on the treeless hills, falling softly upon the Bog of Allen and, farther westward, softly falling into the dark mutinous Shannon waves. It was falling, too, upon every part of the lonely churchyard on the hill where Michael Furey lay buried. It lay thickly drifted on the crooked crosses and headstones, on the spears of the little gate, on the barren thorns. His soul swooned slowly as he heard the snow falling faintly through the universe and faintly falling, like the descent of their last end, upon all the living and the dead.

—JAMES JOYCE, "The Dead"

Tallak taught us the rules: Never step over the gangline. Never trust a snow anchor. Always keep your knife on your belt. Most important, never let go of the sled. If the sled tips over, if it crashes, hold on. Get up if you can. It's nothing to be dragged on your stomach a half mile through the snow. As long as you hold on. If you let go, Tallak said, the dogs and the sled will leave you, and alone in the mountains, on the tundra, you can die. I stuffed my pockets with matches, chocolate, extra mittens, and two headlamps, in case I ever got left behind.

—BLAIR BRAVERMAN, *Welcome to the Goddamn Ice Cube*

I held my girl's hand,
in the deepest parts,
and we walked home, after,
with the snow falling,
but there wasn't much blue
in the drifts or corners:
just white and more white
and the sound track so dead
you could almost imagine
the trees were talking.

—ADRIEN STOUTENBERG, from "Reel One"

The temperature on the moon at night can dip to nearly -250°F.

The temperature of deep space is -454°F, which is about 2.728 degrees above absolute zero, or 0°K (Kelvin).

Absolute zero is the point where no more heat can be removed from a system. In classical kinetic theory, there should be no movement of individual molecules at absolute zero, but experimental evidence shows such is not the case. At the moment, scientists believe that it is not possible to reach absolute zero.

Freeze

Ever since Indo-European preus and Latin *pruina* (hoarfrost), this word has suffered only one mishap, at the hands of the police. Already on record, Japanese visitors to the US have been ordered by the police to "freeze," and have not known how to do it—hands up and keep still. As a cautionary imperative, it is no doubt no more fatuous than melt or pray, but it has done more damage. How often the concisions of a crude elite elude rational folk not even on the run. In one infamous English murder case, a young man named Bentley went to the hang-man for having told an accomplice on a roof, "Let him have it," meaning either *Shoot the cop* or *Let the cop have the gun*. The jury opted for the former, and the judge went along. Bentley's family kept trying to clear his name, but could not. If his death depended on a quibble in interpretation, think what a foreigner, commanded to halt, would make of the colloquial "Hold it." Hold what?

—PAUL WEST, *The Secret Lives of Words*

Lord Kelvin was born as William Thompson in Belfast, Ireland in 1824. He first defined the absolute temperature scale in 1847, which was subsequently named after him, as was the Kelvinator refrigerator. When he died, he was buried next to Isaac Newton in Westminster Abbey.

In Nome, Alaska, glacial melt is depositing tons of gold onto the bottom of the Bering Sea, drawing North, once more, gold seekers willing to work in the cold and difficult conditions.

... when he makes water his urine is congealed ice ...

—SHAKESPEARE, *Measure for Measure*, 3.2.102

Lord Kelvin announced in 1895 that "Heavier-than-air flying machines are impossible."

Ice in flight is bad news. It destroys the smooth flow of air, increasing drag while decreasing the ability of the airfoil to create lift. The actual weight of ice on an airplane is insignificant when compared to the airflow disruption it causes. As power is added to compensate for the additional drag and the nose is lifted to maintain altitude, the angle of attack is increased, allowing the underside of the wings and fuselage to accumulate additional ice. Ice accumulates on every exposed frontal surface of the airplane—not just on the wings, propeller, and windshield, but also on the antennas, vents, intakes, and cowlings. It builds in flight where no heat or boots can reach it. It can cause antennas to vibrate so severely that they break. In moderate to severe conditions, a light aircraft can become so iced up that continued flight is impossible. The airplane may stall at much higher speeds and lower angles of attack than normal. It can roll or pitch uncontrollably, and recovery might be impossible.

—AOPA AIR SAFETY FOUNDATION, https://www.aopa.org/training-and-safety/pic-archive/operations/winter-flying-(2)

Three times the size of Texas and lying 72 degrees North, Greenland is essentially an iced island with 0 percent arable land and no permanent crops. Its resources are minerals, fish, seals, and whales. Vikings reached the island in the tenth century from Iceland. Danish colonization began in the eighteenth century, and Greenland was made an integral part of Denmark in 1953.

I grew up in Rochester, so I should probably know what an ice storm is, but what has just happened here is brand-new in my experience. On the night of March 3, a Sunday, some sort of strange, gentle, superchilled rain came down over a large part of western New York. It coated every power line with a perfect cylinder of ice nearly an inch thick, from which ideally spaced icicles, like the tines on a soil rake, descended, all exactly the same length. The freeze held through early Tuesday. On Monday, if you looked out any window for a few minutes, you were certain to see, against a background of glittering Ace combs, the bough of a tree come crashing down. There was no wind, nor had there been any the first night. It seemed more a demonstration of the patient principles of candlemaking than a storm. At a distance, the ice effects were white, but when you drew close enough to a tree to be surrounded by the continual worry-bead crackling of its fretwork, and stood there, ready to duck at any moment, you saw that it had become incorporated into a clear and disturbingly clinical arrangement of pristine pipettes and test tubes, each holding a one-natural element of the organism—a bud, a twig, one of those perky citizens you had been counting on to function as usual in a few months—in an elaborate cryopharmaceutical experiment.

I drove down the streets today (Tuesday) feeling at times that it was all very familiar, that Ansel Adams calendars had prepared me for this, but then, jumping out again and again from the arty, grainy black-and-white photography that slowly moved past was a sudden apricot-colored splash of discomfort where a bough had torn itself free or a trunk had split in half. On the streets I've seen, half of the good big trees are ravaged. The younger ones, planted about twenty years ago to replace all the elms, are especially painful broken sights. Conifers did somewhat better than deciduous trees. The tall weeping spruce next to our house weeps more than usual but has lost no limbs. Stuck up high in it before the storm was a plastic dragon kite, which we had repeatedly tried and failed to extricate; the morning after the storm, it lay in two pieces on the grass. My wife said, "Well, at least something good has come out of this." I said yes, it was like bombing all of Iraq to get rid of Saddam Hussein. She reminded me of Frost's poem about whether the world will end in fire or in ice. And if we hadn't just flown more than a hundred thousand sorties over a distant place, I would give in more to grief about all these trees, but in the face of that devastation this sort of rare and unmalicious natural catastrophe, in which nobody dies, and some leftovers spoil in some refrigerators, and people go out on tentative camera expeditions to pass the time until the cable TV comes on again, makes me think that we over here have gotten off very easily. We deserve at least this much ice after that much fire. Many of the trees will grow back, after all, as from a bad pruning. As they thaw now, the water is hearteningly visible, hurrying along the bark underneath the ice layer, like blood.

—NICHOLSON BAKER, "Ice Storm"

When the winter chrysanthemums go,
there's nothing to write about
but radishes.

—BASHO (1644–1694)

A tethered horse
snow
in both stirrups.

—BUSON (1716–1783)

Writing shit about new snow
for the rich
is not art.

—ISSA (1763–1827)

Twelve-year-olds suck ice
outside a shop hawking
plantains and child-size caskets.

—DONALD ANDERSON (1946–)

Mush yourself to the North or South Pole, either icecap. Chip from
a glacier for your drink. (You have brought, certainly, a tumbler?)
Add gin, room-temperature tonic. The air escaping the cleaving ice is
unpolluted. Give pause and ear to the whisper—vaunted—of the ages:
a gassy tale 10,000 years told. *Inhale.* You could ask a question now,
but what? Is what you feel in this unadministered place a renunciation
of your ambitions? It may be you must admit to no confusion here on
this lidded retina of the planet, to no sound of wind-chattered flowers,
grass. Why is it you feel no sacrifice, profundum, gain, loss? Will you
find it simpler mushing in than out (as to and from a woman)? Or is
this about blunt waiting: the shift, in the cold, from one foot to the
other, grubbing for lore, instruction? What is it you have for exchange:
fur, soy, clams, fruit? a strong dog's lung? some other untried church?

—DONALD ANDERSON, "Scaling Ice"

Entr'acte

One Christmas Eve at the radar site, one of our senior NCOs, drunk and
stripped bare, armed himself with a rifle and the warning that any officer to
follow to retrieve him would be shot dead for the effort. Then out Travis went
into the Alaskan night. It was one or two in the morning, and windchills were
flirting with 60 below. In the North, Jack Frost hardly nips at you: your toes,
your nose. In the North, real cold will eat you whole. It will, as my father would
say, *outmuscle* you, basing this on his father's—my grandfather's—stories of the
Yukon. We waited until "Doc," our medic, said that we really couldn't wait much
longer. We found Travis headfirst in a drift. He had dropped the rifle. "Doc" kept
him hydrated, blanketed, and sedated until a C-130 could press in through the
weather. Travis was medically discharged and returned, as we understood it, to
his hometown: Murfreesboro, Tennessee. Travis had run the little crafts center
at the radar site where most of us, over the darkened winter, created kiln-
finished ceramic chess sets, wallets, and hand-tooled belts. Travis claimed that
his best childhood pal had been none other than Glen Campbell, the future

country and western star—the "Rhinestone Cowboy." I have always hoped this was true. What I knew was that "Tennessee" Travis hated what he called *DEESE DEPLORABLE VINTERS*! Why he would announce this in what amounted to a Transylvanian or Jewish accent, nobody had a clue.

> Ice shouldn't look so pretty hung from trees
> whose limbs are going to fall on power lines
> and cut me off from light and heat for days.
> Ice ought to be as black and dull as asphalt,
> smearing what it clings to like spilt oil.
> It shouldn't hang from sagging wires like tinsel
> cheering up the dreary winter landscape
> six weeks after Christmas. Sidewalk ice
> should not look like a gleaming coat of varnish
> brushed on to brighten up the dull gray concrete.
> Ice shouldn't turn my neighbor's rusty Olds
> into a fairy coach encased in glass.
> If ice were only beautiful and useless
> as ballet is, as poetry once was,
> I'd praise it as the only luxury
> that people with no money could afford.
> No half week's pay's required to see this ice show.
> All seats are free, with equally good views—
> though I prefer my city-country view,
> where Frost-like glittering maples fill one window
> while iced up streets and sidewalks fill the other.
> Subtract a hundred years and I'd be thrilled
> to jot down notes for "Birches" by my wood stove
> with pen and paper lit with oil lamp light.
> But, like a deep sea diver, I'm attached
> by lifelines that supply me all my needs,
> even letters—pixels on a screen
> that will go blank when my electric's cut,
> (Which will be soon—my lights are flickering.)
> When those wires snap I'll shiver in the dark,
> cursing beauty that I ought to praise

in blank verse lines so eloquent that they're archived
in the 22nd Century Norton Online Anthology.
But somewhere in my desk drawer there's a Bic
with ink enough to write a page or two,
before sun sets, of this mixed review
of Nature's Ice Show—a literal smash hit
as cars out front slide through the intersection
and branches out back shatter on the ice crust.

—RICHARD CECIL,
"One Hundredth Anniversary Edition of 'Birches'"

He giveth snow like wool: He scattereth the hoarfrost like ashes. He
casteth forth His ice like morsels: who can stand before His cold?

—PSALM 147:16–17

At times the sky cleared completely, and sun on the snow was
blinding. As evening fell, the shadows became steel blue, yet the sun
on the horizon was a tomato red. The condition of almost all soldiers,
not just the wounded, was terrible. They limped on frost-bitten feet,
their lips were cracked right open from frost, their faces had a waxen
quality, as if their lives were already slipping away. Exhausted men
slumped to the snow and never rose again. Those in need of more
clothes stripped corpses of clothing as soon as they could after the
moment of death. Once a body froze, it became impossible to undress.

—ANTONY BEEVOR, *Stalingrad*

Soviet divisions were not far behind. 'It is severely cold,' Grossman
noted as he accompanied the advancing troops. 'Snow and the freezing air ice up your nostrils. Your teeth ache. There are frozen Germans, their bodies undamaged, along the road we follow. It wasn't us
who killed them. The cold did. They have bad boots and bad coats.
Their tunics are thin and look like paper. . . . There are footprints all
over the snow. They tell us how the Germans withdrew from the vil-
lages along the roads, and from the roads into the ravines, throwing
their arms away.' Erich Weinart, with another unit, observed crows
circling, then landing, to peck out the eyes of corpses.

—Ibid.

The death rate in the so-called hospitals was terrifying. The tunnel system in the Tsaritsa gorge, redesignated 'Prisoner of War Hospital No. 1', remained the largest and most horrific, only because there were no buildings left offering any protection against the cold. The walls ran with water, the air was little more than a foul, sickly recycling of human breath, with so little oxygen left that the few primitive oil lamps, fashioned from tin, flickered and died constantly, leaving the tunnels dark. Each gallery was not much wider than casualties lying side by side on the damp beaten earth of the tunnel floor, so it was difficult, in the gloom, not to step or trip on feet suffering from frost-bite, provoking a hoarse shriek of pain. Many of the frostbite victims died of gangrene, because the surgeons could not cope. Whether they would have survived amputation in their weakened state and without anesthetic is another matter.

—Ibid.

Ay, but to die, and go we know not where,
To lie in cold obstruction and to rot,
This sensible warm motion to become
A kneaded clod; and the delighted spirit
To bathe in fiery floods, or to reside
In thrilling region of thick-ribbèd ice,
To be imprisoned in the viewless winds,
And blown with restless violence round about
The pendant world; or to be worse than worst
Of that lawless and incertain thought
Imagine howling, 'tis too horrible.
The weariest and most loathèd worldly life
That age, ache, penury, and imprisonment
Can lay on nature is a paradise
To what we fear of death.
—SHAKESPEARE, *Measure for Measure*, 3.1.118–132

In skating over thin ice, our safety is in our speed.
—RALPH WALDO EMERSON

ice bag, ice pack, pack ice, ice pick, ice axe, ice-free, ice show, ice needle, ice storm, ice fog, ice skate, ice hockey, ice bucket, ice cream, ice plant, ice milk, ice blue, ice house, ice chest, ice cave, icebreaker, ice scraper, icemaker, Iceland, icebox, ice bath, icefall, dry ice . . .

ice cap: an extensive dome-shaped or plate-like perennial cover of ice and snow that spreads out from a center and covers a large area, especially of land.

ice foot: a belt or ledge of ice that forms along the shoreline in Arctic regions.

ice barrier: a section of the Antarctic ice shelf that extends beyond the coastline, resting partly on the ocean floor.

ice sheet: a large body of glacier ice spreading in several or all directions from a center; a continental glacier, e.g., Greenland.

ice fall: the part of a glacier resembling a frozen waterfall that flows down a steep slope; an avalanche of ice.

ice field: a large level expanse of floating ice that is more than eight kilometers (five miles) in its greatest dimension.

ice floe: a flat expanse of floating ice that is smaller than an ice field.

iceberg: a massive floating body of ice broken from a glacier. Only about ten percent of its mass is above the surface of the water; a cold, aloof person.

ice, thin.

The Petermann Glacier ice island, an ice chunk four times the size of Manhattan, has broken off a glacier in Greenland. This iceberg contains enough freshwater to keep the Hudson River flowing for two years. A US congressional wit has suggested the iceberg might serve as a *temporary* home for climate-change skeptics.

Ice beer is a higher-alcohol beer produced by chilling below 32 degrees Fahrenheit (0 degrees Celsius) and filtering out the ice crystals that form.

ice cube: a modern convenience.

Slang

ice man: benignly, a person who delivers ice; less benignly, a hired
     killer.

ice: jewelry in general; diamonds in particular.

iced: benignly, cold, or in the case of a place-kicker, delayed; less
benignly, dead.

ice: methamphetamines in the form of visible crystals.

iced: as pertains to a drink or a cake.

ice: also, cocaine, crack cocaine, and PCP.

ice cream habit: use of drugs.

ice cube: crack cocaine.

icing: also cocaine.

snow: also cocaine.

skiing: doing cocaine.

igloo: methylenedioxymethamphetamine.

Ice is actually a mineral, di-hydrogen oxide. Minerals are chemically
homogeneous, with an organized structure of natural, inorganic origin. Ice fits this description perfectly: it has a consistent recipe (H_2O);
it is organized in a symmetrical hexagonal structure; it is formed
naturally and inorganically. Interestingly, mineralogists do not think
of water as a mineral, because it does not have the crystal structure
of ice. While it may also feel strange to think of ice as a rock, it really
is—though it is made up of a single type of molecule, and most rocks
combine different elements. This explains why many geologists are
interested in ice, one of the very few known rocks able to float.[*]

—PAULINE COUTURE, *Ice*

Throughout Earth's history, there have been times when conditions
have proven alternately warmer and colder than they are today. There
have also been several periods, called ice ages, during which large
portions of the planet's surface have been covered with ice. Ice ages,

[*] Pumice, which is forged in molten lava, with trapped air bubbles, also floats.

or glacial periods, have occurred at irregular intervals for more than 2.3 billion years.

The Precambrian period (from 4.6 billion years ago to 570 million years ago) had four ice ages. The first occurred somewhere between 2.7 billion and 2.3 billion years ago. Then the Earth warmed up and was free of ice for almost a billion years. The second ice age occurred between 950 and 890 million years ago; the third between 820 and 730 million years ago; and the fourth between 640 and 580 million years ago. In each case, some area of the Earth was iced over for about 100 million years.

The Mesozoic Era (225 to 65 million years ago) experienced a number of temperature swings, ending in an ice age. This resulted in the extinction of about 70 percent of the living species on Earth, including the dinosaurs. That ice age may have been a result of a collision between an asteroid with the Earth, which created a dust cloud that blocked out the sun.

From 2.4 million years ago to 11,000 years ago, the period known as the "Great Ice Age" (which most people think of as *the* ice age), there were two dozen times when the global temperature plummeted. And for seven different intervals over the last 1.6 million years, up to 32 percent of the Earth's surface has been covered with ice. Scientists estimate that throughout this time period, new ice ages have started about every 100,000 years and have been separated by warmer, interglacial periods, each lasting at least 10,000 years.

The most recent glacial period peaked between 22,000 and 18,000 years ago. At its height, about 27 percent of the world's present land area was covered by ice (compared with 10.4 percent today). There were glaciers up to 10,000 feet (3,050 meters) thick over most of North America, northern Europe, and northern Asia, as well as the southern portions of South America, Australia, and Africa. In North America, the ice covered Canada and moved southward to New Jersey; in the Midwest, it reached as far south as St. Louis. Small glaciers and ice caps also covered the western mountains.

The large ice sheets locked up a lot of water. The sea level fell about 450 feet (137 meters) below what it is today and exposed large areas of land that are currently submerged, such as the Bering land bridge, which connected the eastern tip of Siberia with the western tip of Alaska.

The glaciers' effect on the United States can still be seen. The drainage of the Ohio River and the position of the Great Lakes were shaped by the glaciers. The rich soil of the Midwest is mostly glacial in origin. Rainfall in areas south of the glaciers formed large lakes in Utah, Nevada, and California, such as Utah's Great Salt Lake.

This ice age was followed by a warm period, beginning 14,000 years ago. By 8,000 years ago most of the ice had melted, and between 7,000 and 5,000 years ago the world was 3 to 6° Fahrenheit warmer than it is today.

—"When Was the Ice Age?," enotes.com, http://www.enotes.com/science/q-and-a/when-was-ice-age-288142

Tastee Freez

brain freeze

Prestone

The glacier was hard on bodies, both human and canine, and over the weeks it constantly found new ways to wear us down. First it was the sun—stronger than I had ever felt it, reflecting off the ice so that it shone equally from above and below. On my second day, I had developed blistering sunburn on the insides of my nostrils. Dan had assured me that everyone burned their nostrils in the first week. There wasn't much to do about it, he said, except wait until I built up a nostril tan. Besides, it was nothing compared to the armpit burns that resulted from sunlight shining up the sleeves of a T-shirt, or the sunburns that the male dogs got on their balls, in anticipation of which I smoothed on a layer of diaper rash cream—lick-proof sunscreen—several times a day.

—BLAIR BRAVERMAN, *Welcome to the Goddamn Ice Cube*

Should you find yourself, by intention or necessity, grubbing out a snow cave, do not neglect auguring an air hole (at the least an inch in diameter) in the roof above where you'll sleep to escape poisoning by that old standby: carbon monoxide. Snow insulates so well that without the air hole, you'll die. Besides some warmth and illumination, another reason to light a candle is that, like a canary in a coal mine, it will alert you to the lack of breathable air. Throughout the long night, use a stick or ski pole to ensure the air hole does not clog, in the manner you might check the exhaust of your running car in a snow bank.

> Like a piece of ice on a hot stove the poem must ride on its own melting.
> —ROBERT FROST

> Greenland's surface ice cover experienced a broader thaw during a three-day period this month than in nearly four decades of satellite record-keeping, according to three independent satellite measurements analyzed by NASA and university scientists. About half of the surface of Greenland's ice sheet melts on average each summer. But between July 11 and 13, roughly 97 percent of the sheet experienced some thawing.
> —THE DENVER POST, July 25, 2012

On February 5, 2012, Russian scientists, following two years of drilling, broke through some two miles of Antarctica ice to probe Lake Vostok, a sub-glacial body of water that has been locked in darkness for twenty million years. Scientists trust that this expanse (the size of Lake Ontario) will allow the study of microbial life forms that existed before the most recent ice age. The lake has been kept from freezing by the blanket of two-mile thick ice and the geothermal energy beneath. Contrarily, temperatures at the surface of the Vostok Station have registered the coldest ever recorded on Earth, dipping to -128°F. Vostok Station is located eight hundred miles east of the South Pole.

1870. The US Weather Bureau is established. In 1970, the agency is renamed the National Weather Service.

> Rumors have been spreading that the National Weather Service (NWS) is shedding itself of wind chill values, which describe how cold it feels with wind factored in. The truth is NWS offices in the

Dakotas and Minnesota are *not* eliminating wind chills; they are changing the way they warn the public about extreme, dangerous cold. In the past, the NWS issued Wind Chill Watches and Warnings when wind chill values were expected to fall below a certain level. The problem the NWS encountered was when extreme cold was predicted in the absence of wind, it did not have a warning product that could be issued. "On clear nights when the wind goes calm and we have deep snow cover, that is when we get our coldest temperatures in North Dakota. It can get down to minus 30 or minus 40," said John Paul Martin, warning coordination meteorologist with the National Weather Service office in Bismarck, North Dakota. "We didn't have a product for a warning for that," Martin continued. "Now we're using Extreme Cold Watches and Warnings instead, which are all-encompassing. If it's 30 below or colder, either air temperature or wind chill, we'll issue a warning." The NWS will still include wind chills in their reports and daily forecasts. However, in cases when dangerous cold is predicted, they will issue an Extreme Cold Watch or Warning rather than a Wind Chill Watch, Warning or Advisory.

—see http://www.grandforksherald.com/content/national-weather-service-plans-issue-extreme-cold-warnings-2

"To me, as a meteorologist, if it's 30 below, wind chill or air temperature, it's dangerously cold, and a warning should be issued," Martin stated. "People need to be aware of it." Another issue the Dakotas and Minnesota face is that they are some of the coldest states in the country and deal with subzero cold quite frequently throughout the winter. "The problem is we tend to issue a lot of wind chill advisories," explained Chris Franks, meteorologist with the National Weather Service Office of the Twin Cities in Minnesota. "The goal is to reduce the number of 'cold headlines' we have out and to make the product stronger so that if we do issue a warning, it will be for exceptional events . . . the most dangerous."

—Ibid.

For eight Antarctic summers, I lived and worked in the starkest landscape on Earth. I slept on a mass of ice larger than India and China combined, woke up flying over glaciers that overflow mile-high mountains. . . . Antarctic ice is three miles deep in places. The growth of sea ice around the continent each winter is the greatest seasonal event on the planet, while the largest year-round terrestrial animal is a centimeter-long wingless midge, and plant life is confined to a few stony pockets within the 0.4 percent of Antarctica not covered in permanent snow or ice. Ninety percent of the planet's ice sits in Antarctic ice caps so hostile to human life that nobody stood on them until the twentieth century. The existence of Antarctica, the only significant realm actually discovered by colonizing Europeans, was not proved until 1820, and the continent's coastline was not fully sketched until the late 1950s.

—JASON C. ANTHONY, *HOOSH: Roast Penguin, Scurvy Day, and Other Stories of Antarctic Cuisine*

Entr'acte

The summer before I left Alaska, the old Indian walked me along the river to Gold Camp, five or six miles upstream. He'd hiked in from Koyukuk to visit. He'd brought two dogs and the fish to feed them. The huskies stayed within range, like sheepdogs.

On our way to the camp, we steered clear of the banked gravel and walked on tundra, which gave with our steps and released clouds of mosquitoes and gnats. I wore long sleeves and a cap with netting. When I sprayed my hands with repellant, the old Indian closed his mouth and covered his nose.

He had pointed out the dredged banks of the Indian River. He told me about the hydraulic mining before the war. He meant, he said, "The World War: Number Two." He extended an arm to wave it across the screen of the sky. "Over-digging," he said, toeing the bleached gravel, which stretched out of sight, both banks. "Water mining strips living from earth. Pollutes watersheds. Kills fish."

In Gold Camp, the dredgers were rusted, bear had ransacked the cabins, and returning soldiers had had better offers in the lower states. We found swollen cans of tomatoes the bears had missed, old newspapers, and mattresses full of thin snakes and rodents. From the scat, we could see that even moose had sometimes sought shelter. The old Indian talked about gold and arsenic and spoiled water. He smelled like salmon.

Before we headed back to the radar site, we walked along the river. What I thought to be a glacier canopied the water at a bend. I asked how old the ice was.

"Gone by August," the Indian said.

The ice—blue as sky—towered above. We stood where the river tunneled through, tempted by the microclimate. Following us from the 22-hour summer sun, the mosquitoes and gnats joined black flies in the misty cool. It was as if we were standing inside a bell.

—DONALD ANDERSON, *Gathering Noise from my Life*

The report from Washington, on the day before The Schoolchildren's Blizzard, predicted that Kansas and Nebraska could expect "fair weather, followed by snow, brisk to high southerly winds gradually diminishing in force, becoming westerly and warmer, followed by colder."

—RON HANSEN, "Wickedness"

The origin of the word "blizzard" is obscure. It first appeared in the United States 139 years ago and its first use was possibly on March 14, 1870, to describe a storm that produced heavy snow and high winds in Minnesota. Technically, a blizzard is an intense winter storm with sustained winds 35 mph or higher and sufficient falling and/or blowing snow to reduce visibility below one-quarter mile for at least three hours. Colloquially, the term is loosely applied to any heavy snowstorm. Blizzard conditions figured prominently in one of the worst-ever U.S. winter storms, the "Storm of the Century" that raked the eastern quarter of the nation on March 13–14, 1993, impacting 100 million people in 26 states from Florida to Maine, producing a 12-foot storm surge on Florida's west coast and 20–40 inches of snow and whiteouts from Atlanta to Maine.

—"Blizzards and the 'Storm of the Century,'" *The Chicago Tribune*, March 14, 2005, http://articles.chicagotribune.com/2005-03-14/news/0503140088_1_snow-and-high-winds-storm-blizzard

Snegurochka (Снегу́рочка), or The Snow Maiden, is a character in Russian fairy tales. In one version, she is the daughter of Spring and Frost, and yearns for the companionship of humans. She feels attracted to a shepherd, but her heart is unable to know full love. Her mother, taking pity, affords her this ability, but as soon as Snegurochka falls in love, her heart warms and she melts.

> There must have been some magic in that old silk hat they found,
> For when they placed it on his head, he began to dance around!
> —"Frosty the Snowman"

Jack Frost is a sprite, the personification of winter weather, a variant of "Old Man Winter." In Viking lore, he is known as *Jokul Frosti* or *icicle frost*. It is he who leaves the fernlike crystal patterns on winter windows and nips the extremities in the cold. He has been depicted with paint brush and bucket, coloring the autumnal foliage. If provoked, Jack Frost will kill his detractors by blanketing them with snow.

> I am tired of fighting. Our chiefs are killed. Looking Glass is dead. Toohulhulsote is dead. The old men are all dead. It is the young men who say yes or no. He who led the young men is dead.
>
> It is cold and we have no blankets. The little children are freezing to death. My people, some of them, have run away to the hills and have no blankets, no food. No one knows where they are—perhaps freezing to death. I want to have time to look for my children and see how many I can find. Maybe I shall find them among the dead.
>
> Hear me, my chiefs. I am tired. My heart is sick and sad. From where the sun now stands, I will fight no more forever.
> —Surrender Speech by CHIEF JOSEPH OF THE NEZ PERCE

February 2012. Freezing temperatures across Europe prompted Dutch optimism for the first time in fifteen years that *Elfstedentocht* would be staged. *Elfstedentocht*, or "The 11 Cities Tour," is a 125-mile ice-skating marathon across the frozen canals and lakes in the Dutch province of Friesland, northeast of Amsterdam, beginning and ending in the Frisian capital, Leeuwarden. After the first *Elfstedentocht*, in 1909, conditions have only been icy enough to support the race fifteen times, the last being in 1997. That year there were two hundred contestants, with an additional sixteen thousand skaters following leisurely. The race attracted over two million spectators. Requiring a minimum of fifteen centimeters of ice throughout, the race this present year was, at the last minute, nixed.

> Antarctic exploration of the early twentieth century was unlike exploration of anywhere else on earth. No dangerous beasts or savage natives barred the pioneering explorer's way. Here, with wind speeds up to nearly 200 miles an hour and temperatures as extreme as -100°Fahrenheit, the essential competitions were pure and uncomplicated, being between man and the unfettered force of raw Nature, and man and the limits of his own endurance. Antarctica was also unique in being a place that was genuinely discovered by its explorers. No indigenous peoples had been living there all along, and the men who set foot on the continent during this age could authentically claim to have been where no member of humankind had ever cast a shadow.
>
> —CAROLINE ALEXANDER, *The Endurance: Shackleton's Legendary Antarctic Expedition*

The legend of Hans Brinker, the brave Dutch boy with the Silver Skates, who supposedly stuffed his finger into the dyke to prevent disaster, was actually a literary invention by the American writer Mary Elizabeth Mapes Dodge (1831–1905), who was born in New York.

The clap skate (also called clapskates, slap skates, slapskates, from the Dutch *klapschaats*) is a type of ice skate used in speed skating. Unlike a traditional skate where the blade is rigidly fixed to the boot, a clap skate has the blade attached to the boot by way of a hinge at the toe. This arrangement keeps the blade in contact with the ice longer, as the ankle can now be extended toward the end of the stroke. The setup allows as well for more natural movement, and distributes the energy of the leg more efficiently. During the 1996–1997 season, the Dutch women's team used clap skates with immediate success. The rest of the skating world soon followed, causing a torrent of world records in the following seasons.

Blackborow's condition had become so grave that Macklin and

McIlroy, who were closely monitoring him, had braced themselves for the possibility of having to amputate his feet. By June his right foot seemed to be on the mend, but the toes of the left foot had become gangrenous and needed to be removed. Requiring a temperature high enough to vaporize their scant supply of chloroform, they waited for the first mild day to perform the operation.

On June 15, all hands except for Wild, Hurley, How, and other invalids were sent outside while the Snuggery was converted into an operating theater. A platform of food boxes covered with blankets served as an operating table, and Hurley stoked the bogie stove with penguin skins, eventually raising the temperature to 79°. The few surgical instruments were boiled in the hoosh pot. Macklin and McIlroy stripped to their undershirts, the cleanest layer of clothing they possessed. While Macklin administered the anaesthetic, McIlroy performed the surgery.

—CAROLINE ALEXANDER,
The Endurance: Shackleton's Legendary Antarctic Expedition

The freezing hell of Stalingrad will claim more than 700,000 lives—Germans and Russians, soldiers and civilians—during the winter of 1942. Among the trapped Germans there is evidence of cannibalism. As the "Noose of Steel" tightens, some Russian soldiers admit to pity as they themselves were well dressed against the cold and well fed. Of the 300,000 German troops in the noose, two-thirds will die. One million Russians had flanked the German 6th Army just as winter arrived with the vengeance of -20 degree temperatures. Of the 33,000 Germans to survive for surrender and work camps, one-third died while traversing the Soviet Union in unheated train cars, in which they were fed but every three days. The survivors worked in remote factories and mines (coal and uranium) where temperatures often dipped to -75°F. Only 6,000 German soldiers ever make it back to Germany. In other words, in Stalingrad, 50 grams of bread a day, coupled with minus 50°F weather, claims more victims than enemy action. There wasn't enough clothing, fire, dead horses, dogs, or cats to keep them alive.

—ANTONY BEEVOR, *Stalingrad*

Every sledging party that dared the Antarctic interior was severely dehydrated, largely because Antarctic air is the atmospheric equivalent of toast. Antarctica is the driest continent, a great polar desert where every breath exhaled carries moisture from the lungs that cannot be replaced except by drinking melted snow or ice. Fuel was limited, however, and explorers could only rehydrate at meals. Eating snow or ice directly burns more energy in melting and warming it to body temperature than the drink provides. Sledgers did not often carry water containers, which were too heavy and too quickly frozen. . . . This chronic dehydration occurred as the men marched over, fell into, or were buried by the largest mass of frozen water on Earth.

—JASON C. ANTHONY, *HOOSH: Roast Penguin, Scurvy Day, and Other Stories of Antarctic Cuisine*

Be thou chaste as ice, as pure as snow, thou shalt not escape calumny.
—SHAKESPEARE, *Hamlet*, 3.1.137

2009. Dr. Jerri Nielsen FitzGerald, who diagnosed and treated her own breast cancer before a dramatic rescue from the South Pole, has died. She was 57. Her cancer had been in remission. FitzGerald was the only doctor among 41 staff at the National Science Foundation's Amundsen-Scott South Pole Station in winter 1999 when she discovered a lump in her breast. At first, she didn't tell anyone, but the burden became too much to bear. "I got really sick," she told The Associated Press in a 2003 interview. "I had great big lymph nodes under my arm. I thought I would die." Rescue was out of the question. Because of the extreme weather conditions, the station is closed to the outside world for the winter. She had no choice but to treat the disease herself, with help from colleagues she trained to care for her and U.S.-based doctors she stayed in touch with via satellite e-mail. She performed a biopsy on herself with the help of staff. A machinist helped her with her IV and test slides, and a welder helped with chemotherapy. She treated herself with anti-cancer drugs delivered during a mid-July airdrop by a U.S. Air Force plane in blackout, freezing conditions. She was air-lifted by the Air National Guard in October, one of the earliest flights ever into the station when it became warm enough—58 degrees below zero—to make the risky flight. After multiple surgeries in the U.S., including a mastectomy, the cancer went into remission until 2005.

—"Jerri FitzGerald, Who Treated Herself at South Pole, Dies at 57," *The New York Times*, June 25, 2005, http://www.nytimes.com/2009/06/25/us/25nielsen.html

Dark spruce forest frowned on either side of the frozen waterway. The trees had been stripped by a recent wind of their white covering of frost, and they seemed to lean toward each other, black and ominous, in the fading light. A vast silence reigned over the land. The land itself was a desolation, lifeless, without movement, so lone and cold that the spirit of it was not even that of sadness. There was a hint in it of laughter, but of a laughter more terrible than any sadness—a laughter that was mirthless as the smile of the Sphinx, a laughter cold as the frost and partaking of the grimness of infallibility. It was the masterful and incommunicable wisdom of eternity laughing at the futility of life and the effort of life. It was the Wild, the savage, frozen-hearted Northland Wild.

—JACK LONDON, *White Fang*

I'd gained nearly seven hundred feet of altitude since stepping off the hanging glacier, all of it on crampon front points and the picks of my axes. The ribbon of frozen meltwater had ended three hundred feet up and was followed by a crumbly armor of frost feathers. Though just barely substantial enough to support body weight, the rime was plastered over the rock to a thickness of two or three feet, so I kept plugging upward. The wall, however, had been growing imperceptibly steeper; and as it did so, the frost feathers became thinner. I'd fallen into a slow, hypnotic rhythm—swing, swing; kick, kick; swing, swing; kick, kick—when my left ice ax slammed into a slab of diorite a few inches beneath the rime.

I tried left, then right, but kept striking rock. The frost feathers holding me up, it became apparent, were maybe five inches thick and had the structural integrity of stale corn bread. Below was thirty-seven hundred feet of air, and I was balanced on a house of cards. The sour taste of panic rose in my throat. My eyesight blurred, I began to hyperventilate, my calves started to shake. I shuffled a few feet farther to the right, hoping to find thicker ice, but managed only to bend an ice ax on the rock.

—JON KRAKAUER, *Into the Wild*

But we little know until tried how much of the uncontrollable there is in us, urging across glaciers and torrents, and up dangerous heights, let the judgement forbid as it may.

—JOHN MUIR, *The Mountains of California*

Awkwardly, stiff with fear, I started working my way back down. The rime gradually thickened. After descending about eighty feet, I got back on reasonably solid ground. I stopped for a long time to let my nerves settle, then leaned back from my tools and stared up at the face above, searching for a hint of solid ice, for some variation in the underlying strata, for anything that would allow passage over the frosted slabs. I looked until my neck ached, but nothing appeared. The climb was over. The only place to go was down.

—JON KRAKAUER, *Into the Wild*

Rime Ice (as defined by the National Weather Service): An opaque coating of tiny, white, granular ice particles caused by the rapid freezing of super-cooled water droplets on impact with an object. See also clear ice.

Bridge Ices Before Roadway

Scientists in Russia have made a major breakthrough in permafrost research. The team, whose work is based in Siberia, successfully germinated a flower from an ice-age seed which is about 32,000 years old. The scientists found an ancient frozen nest of Arctic ground squirrels 30 metres underground. They took some of the seeds they found in that nest and brought them back to a lab. Their attempts to try to resurrect ice-age life from the permafrost were successful—they produced a small white tundra flower from the seed. The plant is called

Silene stenophylla, which is like an ice-age version of a chickweed—a flower which lives in very dry tundra.

—"Scientists germinate ancient seed, bring back to life ice age flower," Alaska Dispatch, http://www.alaskadispatch.com/article/scientists-germinate-ancient-seed-bring-back-life-ice-age-flower

Léxuwakipa, who was very touchy, felt offended by the people. In revenge she let it snow so much that an enormous mass of ice came to cover the entire Earth. When it eventually began to melt, there was so much water that the Earth became completely flooded.

—YÁMANA CREATION STORY

Ice had formed ahead of them, and it reached all the way to the sky. The people could not cross it. . . . A Raven flew up and struck the ice and cracked it. Coyote said, "These small people can't get across the ice." Another Raven flew up again and cracked the ice again. Coyote said, "Try again, try again." Raven flew up again and broke the ice. The people ran across.

—PAIUTE LEGEND

Madumda made new people from willow wands, taught them to hunt with a bow and arrow, how to make baskets, and how to eat, and then he went home to the north. But these people went bad too, so Madumda sent ice down to kill them all.

—POMO CREATION STORY

North Pole, Alaska (zip code 99705), claiming a population of 1,600 is 1,750 miles south of its namesake, and is a bedroom community for Fairbanks and two military bases fourteen miles away. On December 21, you can expect three hours and forty-two minutes of sunlight. On June 21, twenty-one hours and forty-nine minutes of rays. The annual average of letters to Santa received at the post office numbers one hundred thousand.

Cabin fever is an idiomatic term for a claustrophobic reaction that takes place when a person is isolated and confined, with nothing to do. Symptoms include restlessness, irritability, paranoia, irrational frustration with everyday

38

objects, forgetfulness, laughter, excessive sleeping, distrust of anyone they are with, and an urge to go outside, even into life-threatening weather.

> In all the world there is no desolation more complete than the polar night. It is a return to the Ice Age—no warmth, no life, no movement. Only those who have experienced it can fully appreciate what it means to be without the sun day after day and week after week. Few men unaccustomed to it can fight off its effects altogether, and it has driven some men mad.
>
> —ALFRED LANSING, *Endurance: Shackleton's Incredible Voyage*

The date was October 27, 1915. The name of the ship was *Endurance*. The position was 69°5' South, 51°30' West—deep in the icy wasteland of the Antarctic's treacherous Weddell Sea, just about midway between the South Pole and the nearest known outpost of humanity, some 1,200 miles away. . . .

They were for all practical purposes alone in the frozen Antarctic seas. It had been very nearly a year since they had last been in contact with civilization. Nobody in the outside world knew they were in trouble, much less where they were. They had no radio transmitter with which to notify any would-be rescuers, and it is doubtful that any rescuers could have reached them even if they had been able to broadcast an SOS. It was 1915, and there were no helicopters, no Weasels, no Sno-Cats, no suitable planes.

Thus their plight was naked and terrifying in its simplicity. If they were to get out—they had to get themselves out.

> —Ibid.

> Out of whose womb came the ice?
> And the hoary frost of Heaven
> who hath gendered it?
> The waters are hid as with a stone.
> And the face of the deep is frozen.
>
> —JOB 38:29

The Carpathian campaign of 1915, frequently dubbed the "Stalingrad of the

First World War," pitted the million-man armies of Austria-Hungary and Russia in fierce winter combat that resulted in hundreds of thousands of soldiers succumbing to frostbite and what came to be known as the "White Death."

For Second Army troops, noncombat casualties continued to outnumber battlefield losses. Indicative of this, one unit lost 150 men to enemy fire, 200 to frostbite. Troops were forced to use snow for cover and protection, which increased the likelihood of frostbite. The 27th Infantry Division found itself hampered by up to two meters of snow, while the strength of the XIX Corps dissipated drastically. With no relief in sight, troops became even more demoralized. Suicide remained a viable solution for some troops.

—GRAYDON A. TUNSTALL, *Blood on the Snow:*
The Carpathian Winter War of 1915

Another use of the term "White Death" was the nickname the Red Army accorded to a Finnish sniper who recorded 505 confirmed Russian kills. In temperatures reaching -40°F, camouflaged in white, using an ordinary rifle without telescopic sights, Simo Häyhä carried out his missions. In the spring of 1940, during combat, he was shot in the face by a Russian soldier. The exploding bullet crushed his jaw and blew away his left cheek. He survived and lived until 2002. At war's end he was promoted from corporal to second lieutenant. One of his tricks, he said, was to keep snow in his mouth so that the vapor of his breath would not reveal his position.

—ALLEN F. CHEW, *White Death:*
The Epic of the Soviet-Finnish Winter War

On 23 January we rushed forward in the icy hell of the Carpathian battlefield. We stormed the Uzsok, Verecke and Wyszkov Passes, but on the northern slope of the mountains, the troops encountered a blizzard. . . . Everyday [*sic*] hundreds froze to death. The wounded that were unable to drag themselves forward were left to die. Entire ranks were reduced to tears in the face of the terrible agony.

Each night the 21st Infantry Regiment dug in until the last man was found frozen to death at daybreak. Pack animals could not advance through the deep snow. The men had to carry their own supplies on foot. The soldiers went without food for days. At -25°C, food rations froze solid. For seven days straight, the 43rd Infantry Division battled overpowering Russian troops with no warm food to sustain them. For a full thirty days, not one single man had any shelter.

—COLONEL GEORG VEITH, *Tunstall*

Three million Finns holding out against a nation of 171 million? In the famous battle of "Killer Hill," 32 Finns battled 4,000 Soviet soldiers. Simo Häyhä accomplished his 505 kills in 100 days, before he was shot himself. Just for the record, when he was shot in the face, Häyhä managed to retrieve his rifle and shoot the Russian who had shot him.

—ALLEN F. CHEW, *White Death:*
The Epic of the Soviet-Finnish Winter War

Entr'acte

The winter Alexandra was in her second year of high school and newly possessed of a driving permit, a heater hose in the old Datsun burst. Twenty years, and I'm stuck with that part of that day. Sky looks greasy, pressured for storm. Air like suet. 4 p.m. Cold.

Alex had expected to drive to SAFEWAY, been headed for the wheel. "You think we don't have snow on the roads at school? We had to watch a zillion flicks. Snow flicks. I'm driving."

The slab where I parked the Datsun had been coated from storms, ice fattening for weeks. The night temperature had so steadfastly nicked zero that the mark had become a temperature you could believe in, a location. Like the inside of a bell. The interior of a cathedral.

The slab ice only melted from the heat of a just-parked car. Each morning I backed the Datsun out, I saw the shell restored. In the mornings the ice looked permanent. Now, spreading from beneath the car, was unfreezable muck. The leaked antifreeze had the look of clabber.

"What is it?" Alex asked. "What's that?"

"Don't step in it."

"What is it?"

"Get your feet out of it, Alex. Go in."

"I need lunch. I'm not buying at school." She was back toeing the spill.

"I told you, go in."

"I need stuff."

I sat Alex on the porch. "Sit down." I unlaced my sixteen-year-old's boots, then pulled them, then pointed at the door. She went in.

—DONALD ANDERSON, "Fire Road"

An *Obergefreiter* will be court-martialled because he is supposed to have deliberately allowed both his feet to get frozen. Before they brought him to the medics he told us that following a Russian attack he had saved himself by playing dead. In order to avoid being detected by the enemy, he had spent the entire night in a snowdrift. When another combat unit freed him the next morning during a counter-attack, his feet were two blocks of ice.

—GUNTER K. KOSCHORREK, *Blood Red Snow:
The Memoirs of a German Soldier on the Eastern Front*

Speaking of anthropological canards, no discussion of language and thought would be complete without the Great Eskimo Vocabulary Hoax. Contrary to popular belief, the Eskimos do not have more words for snow than do speakers of English. They do not have four hundred words for snow, as has been estimated in print, or two hundred, or one hundred, or forty-eight, or even nine. One dictionary puts the figure at two. Counting generously, experts can come up with about a dozen, but by such standards English would not be far behind, with *snow, sleet, slush, blizzard, avalanche, hail, hardpack, powder, flurry, dusting,* and a coinage of Boston's WBZ-TV meteorologist Bruce Schwoegler, *snizzling.*

—STEPHEN PINKER, *The Language Instinct*

Among the many depressing things about this credulous transmission and elaboration of a false claim is that even if there were a large number of roots for different snow types in some Arctic language, this would not, objectively, be intellectually interesting; it would be a most mundane and unremarkable fact. Horsebreeders have various names for breeds, sizes, and ages of horses; botanists have names for leaf shapes; interior decorators have names for shades of mauve; printers have different names for fonts (Carlson, Garamond, Helvetica, Times Roman, and so on), naturally enough. . . . Would anyone think of writing about printers the same kind of slop we find written about Eskimos in bad linguistics textbooks?

—GEOFFREY PULLMAN, qtd. in
Stephen Pinker, *The Language Instinct*

Today's Finnish Army keeps 700,000 pairs of skis ready to wax for the precise snow conditions that might accompany an invasion. Beginning at age twenty, every Finnish male must serve two years in the army. Every soldier and every officer learns to ski and fight at the same time.

The Finns remain on the cutting edge of military ski technology. They have developed relatively long, wide, heavy skis, ranging from 210 to 250 cm, and 80 to 100 mm in width. Wider than most alpine skis, these tools of war have no side-cut. This seeming awkward design keeps soldiers always on top of the snow, so they never have to break trail through deep powder while approaching or fleeing an enemy.

The Finns's armored personnel carriers feature two cabs setting on four separate tracks, with a rotating power linkage between. The two halves twist and turn independently through the snow, while the vehicle's narrowness lets it squeeze between trees. Each machine can carry twenty-four men and tow another twenty-four on skis.

—TOM WOLF, *Ice Crusaders: A Memoir of Cold War and Cold Sport*

When Norwegian Bredo Morstoel expired in his native land in 1989, he was packed in dry ice and shipped to the Trans Time cryonics facility in Oakland, California, where he was placed in liquid nitrogen for the next four years, after which he was moved to Nederland, Colorado, to stay with his daughter Aud Morstoel and his grandson Trygve Bauge, both advocates for cryonics hoping to establish a facility of their own. Since arriving, Bredo has remained in Nederland under cold cover, in a Tuff shed, near his grandson's home. When Trygve was deported in the mid-'90s, Bredo's daughter stepped in to take care of keeping her father preserved. Not long after, Aud was evicted for living in a house with no electricity or plumbing and was about to head back to Norway. Worried that her father would thaw without her attention, she spoke to a local reporter, who spoke to the Nederland city council, who passed Section 7–34 of the municipal code regarding the "keeping of bodies." Grandpa Bredo was grandfathered in and allowed to stay. As a consequence of media, the "Frozen Dead Guy" became a worldwide sensation. Responding to a want ad on the Internet in 1995, local Bo Shaffer applied for the one-of-a-kind job and has come to be known as the "Ice Man." Once a month, Shaffer and a team of volunteers deliver 1,600 pounds of dry ice to pack around Bredo in his sarcophagus. Thus, Bredo is kept at a steady -60°F.

—see http://www.businessinsider.com/bo-shaffer-kept-
bredo-morstoels-corpse-frozen-for-18-years-2014-3

Nederland's annual "Frozen Dead Guy Days Festival" arrives at winter's end and is jam-packed each March with activities: a hearse parade, frozen salmon tossing, frozen turkey bowling, a polar-bear plunge into Chipeta Park Pond, and a coffin race. The Bredo family has founded the International Cryonics Institute and Center for Life Extension—*yes, ICICLE*—on the outskirts of town, but the institute's "research" is probably best described as "low-tech," consisting as it has (and does) of a Tuff Shed, a stainless-steel sarcophagus, and the monthly delivery of nearly a ton of dry ice.

After Hall of Fame baseball player Ted Williams died in 2002, his body was transported by private jet to the Alcor Life Extension Foundation in Scottsdale, Arizona. There, Williams's body was separated from his head in a procedure called *neuroseparation*. The head is stored in a steel can filled with liquid nitrogen, and Williams' body stands upright in a nine-foot tall cylindrical steel tank, also filled with liquid nitrogen, both head and body frozen to a surprising -321°F. The procedure, approved by Williams's son and daughter, carried a $136,000 price tag. Alcor claims it is still owed over $100,000, but also denies that its employees mistreated and damaged Williams's head in a bizarre "batting practice" escapade complete with a monkey wrench—all reported in *Frozen*, a tell-all book authored by Larry Johnson, a former executive at Alcor.

In 1991, the oldest frozen body ever discovered was found by two hikers in

44

the Alps. No ordinary glacial mummy, "The Iceman" was determined to be some five thousand years old and the most celebrated cadaver since King Tut. The Iceman is thought to have been about forty-five years old when he met his end on the mountain. Researchers have determined that The Iceman's last full meal, 5,300 years ago, consisted of ibex and deer meat, sloe plums, and wheat bread. Researchers also claim that the man lived a life that included strenuous walking in high altitudes and that he suffered from knee pain.

And no one can tell you why your toe goes numb or why your elbow hurt last week?

It was Dr. Frederick Cook who helped Roald Amundsen, on an early sail to Antarctica, to learn from the more northern Inuits how to keep warm using reindeer hides for clothing and boots. Cook also taught the man destined to be (almost fifteen years later) the first to reach the actual South Pole about the importance of fresh food in keeping any crew healthy. Because there were no fruits or vegetables in the frozen lands they were exploring, men aboard Amundsen's ships were taught to consume raw seal and penguin, both of which contain enough vitamin C to ward off scurvy, that longtime plague of sailors.

> After several days of clear calm, a strong gale arose on July 12 and grew into a full-blown blizzard on July 13. The ship quivered as the pressure ground around her. Wild and Worsley were visiting with Shackleton in his cabin.
> "The wind howled in the rigging," Worsley recalled, "and I couldn't help thinking that it was making just the sort of sound that you would expect a human being to utter if he were in fear of being murdered." In the lulls of the wind, the three men listened to the grinding of ice against the ship's sides. It was now that Shackleton shared what he had known for many months.
> "The ship can't live in this, Skipper," he said, pausing in his restless march up and down the small cabin. "You had better make up your mind that it is only a matter of time. It may be a few months, and it may be only a question of weeks, or even days . . . but what the ice gets, the ice keeps."
> —CAROLINE ALEXANDER, *The Endurance: Shackleton's Legendary Antarctic Expedition.*

Roald Amundsen arrived in Hobart, Australia, the first week of March in 1912, from where he dispatched telegrams announcing his successful expedition to the South Pole the previous December.

On November 12, 1912, an Antarctic search party discovered its objective—the tent of Captain Robert Scott and his two companions half-buried in the snow. Inside, they found the body of Captain Scott wedged between those of his fellow explorers, the flaps of his sleeping bag thrown back, his coat open. His companions, Lieutenant Henry Bowers and Dr. Edward Wilson, lay covered in their sleeping bags, as if dozing. They had been dead for eight months. They were the last members of a five-man team returning to their home base from the Pole.

The team had set out on its final push to the Pole the previous January. They knew they were in a race to be the first to reach their destination. Their competition was the Norwegian expedition led by Amundsen. Amundsen relied on dogs to haul his men and supplies over the frozen Antarctic wasteland. Scott's British team distrusted the use of dogs, preferring horses, but once the horses perished from the extreme conditions, the sleds had to be man-hauled to the Pole and back.

In addition to Captain Scott, Lieutenant Bowers, and Dr. Wilson, two others, Captain Titus Oates and Petty Officer Edgar Evans had made the final push to the Pole. Conditions were appalling: temperatures plummeting to -45°F, nearly impassable terrain, blinding blizzards, or blinding sunshine. On January 16, nearing their objective, Scott and his team made the disheartening discovery that the Norwegians had beaten them to the Pole. In fact, the Norwegians had arrived four weeks earlier on December 14, 1911. Psychologically numbed by the finding, the team pushed on.

—"Doomed Expedition to the South Pole, 1912," EyewitnesstoHistory. com, http://www.eyewitnesstohistory.com/scott.htm

From Scott's Diary:
Thursday morning, January 18— . . . We have just arrived at this tent, 2 miles from our camp, therefore about 1 ½ miles from the Pole. In

the tent we find a record of five Norwegians having been here . . . We carried the Union Jack about ¾ of a mile north with us and left it on a piece of stick as near as we could fix it . . . Well, we have turned our back now on the goal of our ambition and must face our 800 miles of solid dragging—and good-bye to most of the day-dreams!

Friday, March 16 or Saturday 17—Lost track of dates, but think the last correct. Tragedy all along the line. At lunch, the day before yesterday, poor Titus Oates said he couldn't go on; he proposed we should leave him in his sleeping-bag. That we could not do, and we induced him to come on, on the afternoon march. In spite of its awful nature for him he struggled on and we made a few miles. At night he was worse and we knew the end had come. Should this be found I want these facts recorded. Oates's last thoughts were of his Mother, but immediately before he took pride in thinking that his regiment would be pleased with the bold way in which he met his death. We can testify to his bravery. He has borne intense suffering for weeks without complaint, and to the very last was able and willing to discuss outside subjects. He did not—would not—give up hope till the very end. He was a brave soul. This was the end. He slept through the night before last, hoping not to wake; but he woke in the morning—yesterday. It was blowing a blizzard. He said, 'I am just going outside and may be some time.' He went out into the blizzard and we have not seen him since.

Thursday, March 29—Since the 21st we have had a continuous gale from W. S. W. and S. W. We had fuel to make two cups of tea apiece and bare food for two days on the 20th. Every day we have been ready to start for our depot 11 miles away, but outside the door of the tent it remains a scene of whirling drift. I do not think we can hope for any better things now. We shall stick it out to the end, but we are getting weaker, of course, and the end cannot be far. It seems a pity, but I do not think I can write more.

—Ibid.

On January 28, the Russians divided the city into three sectors: the Eleventh Corps was isolated around the tractor plant; the Eight and Fifty-first Corps around an engineering school west of Mamaev Hill;

the remainders of the Fourteenth and Fourth Corps were in the downtown area around the *Univermag.*

At the *Schnellhefter* Block, across from the tractor plant, Dr. Ottmar Kohler had run out of morphine. Wallowing in filth and blood, he operated under flickering lights and in incredible cold. Outside the building, lines of soldiers crowded the entrance, looking for a place to sleep. An officer went to the door and begged them to go away because there was no room, but they said they would wait until morning.

At sunrise, the visitors were still there, huddled together against the below-zero temperature. During the night they had all died from exposure.

—WILLIAM CRAIG, *Enemy at the Gates: The Battle for Stalingrad*

"There was a cruel aftermath to the blizzard," wrote a survivor, "funerals, surgical operations, cripples, fingers with first joints gone, ears without rims, and some like poor Will Moss, who spent the night on the prairie in the shelter of his cutter, and supposed that he had escaped without damage, afterward died of diseases caused by the exposure."

The precise number of the dead was never determined. Estimates published over the years in state histories and local newspapers have ranged from 250 to 500. The southern and eastern part of Dakota Territory suffered the majority of the casualties. Undoubtedly many deaths were never reported from remote outlying districts. Scores died in the weeks after the storm of pneumonia and infections contracted during amputations. For years afterward, at gatherings of any size in Dakota or Nebraska, there would always be people walking on wooden legs or holding fingerless hands behind their backs or hiding missing ears under hats. . . .

—DAVID LASKIN, *The Children's Blizzard*

One winter there was a freezing rain. How beautiful! people said when things outside started to shine with ice. But the freezing rain kept coming. Tree branches glistened like glass. Then broke like glass. Ice thickened on the windows until everything outside blurred. Farmers moved their livestock into the barns, and most animals were safe. But not the pheasants. Their eyes froze shut.

Some farmers went ice-skating down the gravel roads with clubs to harvest the pheasants that sat helplessly in the roadside ditches.

The boys went out into the freezing rain to find pheasants too. They saw dark spots along a fence. Pheasants, all right. Five or six of them. The boys slid their feet along slowly, trying not to break the ice that covered the snow. They slid up close to the pheasants. The pheasants pulled their heads down between their wings. They couldn't tell how easy it was to see them huddled there.

The boys stood still in the icy rain. Their breath came out in slow puffs of steam. The pheasants' breath came out in quick little white puffs. Some of them lifted their heads and turned them from side to side, but they were blindfolded with ice and didn't flush. The boys had not brought clubs, or sacks, or anything but themselves. They stood over the pheasants, turning their own heads, looking at each other, each expecting the other to do something. To pounce on a pheasant, or to yell Bang! Things around them were shining and dripping with icy rain. The barbed-wire fence. The fence posts. The broken stems of grass. Even the grass seeds. The grass seeds looked like little yolks inside gelatin whites. And the pheasants looked like unborn birds glazed in egg white. Ice was hardening on the boys' caps and coats. Soon they would be covered with ice too.

Then one of the boys said, Shh! He was taking off his coat, the thin layer of ice splintering in flakes as he pulled his arms from the sleeves. But the inside of the coat was dry and warm. He covered two of the crouching pheasants with his coat, rounding the back of it over them like a shell. The other boys did the same. They covered all the helpless pheasants. The small gray hens and the larger brown cocks. Now the boys felt the rain soaking through their shirts and freezing. They ran across the slippery fields, unsure of their footing, the ice clinging to their skin as they made their way toward the blurry lights of the house.

—JIM HEYNEN, "What Happened During the Ice Storm"

The fireplace is insatiable, ravenous. It is a beast I must feed steadily to keep its energies alive and leaping. I'm glad the woodpile is stacked high. I'll have to add to it now and then to keep it looking healthy. At the rate the logs go, I'll have to do it for survival.

How did the Indians ever keep warm through these long winters? Between their constant need for firewood and the beavers' need for food and lodging, it's a wonder there were any trees left in these woods by spring.

—EDWARD LUEDERS, *The Clam Lake Papers:*
A Winter in the North Woods

I had come to the Arctic for adventure, but I also carried with me a circular logic: If I could be safe in this land, maybe I could be safe in my own body. If I could protect my body, maybe I could live in this land. So far, though, I was not doing anything right. To start with, I had no wool sweaters, and wool sweaters in the Northland were of nearly religious significance. If it was thirty below and you were cold, it was your own fault because your sweater wasn't wool. Or if your sweater was wool, maybe your shirt wasn't, maybe your bra wasn't. I didn't own any wool at all, so I went to the nearby shop in hopes of finding some.

—BLAIR BRAVERMAN, *Welcome to the Goddamn Ice Cube*

Scalp Hypothermia:
During your chemotherapy, ice packs or similar devices are placed on your head to slow blood flow to your scalp. This way, chemotherapy drugs are less likely to have an effect on your scalp. Studies of scalp hypothermia have found it works somewhat in the majority of people who have tried it. However, the procedure also causes a small risk of cancer recurring in your scalp, as this area doesn't receive the same dose of chemotherapy as the rest of your body. People undergoing scalp hypothermia report feeling uncomfortably cold and having headaches.

—"Chemotherapy and Hair Loss:
What to Expect During Treatment," Mayo Clinic,
http://www.mayoclinic.com/health/hair-loss/CA00037

"In the dead of night." "In the dead of winter." These expressions seem altogether natural, and on the surface, the Clam Lake woods and much of the life that inhabits them support such expressions. The winter landscape seems dead and dispirited; the winter night, doubly so.

Yet the cold is a preserver, and the rampant, wild metabolism of summer in this north-woods country, as I consider it today, is much more often the destroyer. Just as heat is the speeded-up action of molecules and cold is their comparatively calm repose, the frenzy of summer is relieved and complemented by the freeze (the frieze?) of winter. The physical world is composed of molecules, their dance slowed to a frigid minimum in winter. The molecular world does not die but achieves a greater composure. In other words, the decomposition of the organic world is arrested. Slowed down, dormant, apparently inactive, it invites the mind. "The dead of winter" is an inaccurate phrase. "The deliberation of winter" might be better.

—EDWARD LUEDERS, *The Clam Lake Papers:*
A WINTER in the North Woods

Where is the Arctic? An ocean surrounded by continents, and an indistinct geographical zone. The top part—the High Arctic—is a dazzling hinterland where myth and history fuse, a white Mars. . . . The southern limits of the Arctic are movable feasts. Some people—such as the farmers in northern Sweden and Finland who wish to claim EU Arctic farming subsidies—consider that the Arctic Circle at 66°33' constitutes the frontier. In Canada the definition fails, as swaths of typical Arctic territory (permafrost, permanent ice cover, absence of topsoil and significant vegetation, polar bears, all of these) lie well south of the Arctic Circle. Sixty-six degrees simply marks the point at which the sun fails to set at the summer solstice (June 21) or rise at the winter solstice (December 21); climatological and other factors produce divergent conditions at different points on the Circle. The Gulf Stream warms the oceans and the surrounding air to create clement conditions for the subsidy-claiming Scandinavian farmers, and in Finland the viviparous lizard thrives north of the Arctic Circle, whereas parts of the sub-Arctic are colder than anywhere else on earth. The residents of Oymyakon in the Sakha Republic three degrees outside the Arctic Circle once recorded a temperature of -97°F, a level at which trees explode with a sound like gunfire and exhaled breath falls to the ground in a tinkle of crystals.

—SARA WHEELER, *Magnetic North: Notes from the Arctic Circle*

"Colder than a witch's tit," my wife will say when the temperature drops.

Many Siberians refer to western Russia as the materik—the Mainland, which is similar to the way Alaskans think of the Lower 48. But most of Siberia is thousands of miles closer to the capital than Russia's Far East. In terms of sheer remoteness—both geographic and cultural—the Maritime Territory is more like Hawai'i. . . . In Primorye, the seasons collide with equal intensity: winter can bring blizzards and paralyzing cold, and summer will retaliate with typhoons and monsoon rains; three quarters of the region's rainfall occurs during the summer. This tendency toward extremes allow for unlikely juxtapositions and explains why there is no satisfactory name for the region's peculiar ecosystem—one that happens to coincide with the northern limit of the tiger's pan-hemispheric range. It could be argued that this region is not a region at all but a crossroads: many of the aboriginal technologies that are now considered quintessentially North American—tipis, totem poles, bows and arrows, birch bark canoes, dog sleds, and kayak-style paddles—all passed through here first.

—JOHN VAILLANT, *The Tiger:*
A True Story of Vengeance and Survival

Herds of reindeer moved across ice and snow. Slim-shouldered Lapps squatting on Ski-Doos nosed their animals toward an arc of stockades. A man in a corral held a pair of velvet antlers while another jabbed a needle into a damp haunch. I made my way toward the outer palisades, where Lapps beyond working age stoked beech wood fires and gulped from bowls of reindeer broth, their faces masked in musky steam. The first new snow had fallen, and the Harrå Sámi were herding reindeer down to the winter grazing. A livid sun hung on the horizon. Sámi, or Lapps, were the last nomadic people in Europe, and until recently they castrated reindeer at this place by biting off their balls.

—SARA WHEELER, *Magnetic North: Notes from the Arctic Circle*

Entr'acte

In the yard, I stuffed my hands into Alex's boots, then toed them in the snow to clean them. Then I straightened. There was still a good hour of day, but the sky looked brighter than it seemed it should have, as though the day's temperature had altered light. I listened for noise, but the cold had made everything quiet. I took the time just to stand there, shut my eyes. Alex interrupted from the porch. She wanted her boots. I kept the boots on my hands. She shut the door.

The yellow-green pool under the car had worked its way now around a tire, and I watched as if I expected the muck to climb and work the rubber: dog-hungry, brained glop from some raw planet on a search. I so disliked the sight of my carport ice defiled that I called up the clean vision of an Arctic fox I'd spied one winter in Alaska's Brooks Range, just north of the Arctic Circle. It was a healthy male whose coat should have shone fox-silver, yet there he sped, terra cotta against unmarked snow. That spell, lasting but seconds, had remained one of the permanent visions of my life: a beauty I felt in unblessed surrender to the spill of antifreeze beneath my car. In that unvirtuous green (reflecting a decaying undercarriage), I faced what seemed a further and containing vision of the swift destruction of the North after piled centuries of pure extreme fragile life: my red fox, miles upon miles of what had seemed my own owned ice (unearthly white and pale sky-blue unending), perfected storm. And in summer: rock, sun, thaw, rain, hail, illogical flowers, gnats, nervy salmon, caribou, bear. After all the years, my year in the North still counted.

I clapped Alex's boots as much for noise as to jar snow, then tossed them at the porch. Alex cracked open the door, took them. I raised the car's hood, peered in, dropped the hood. Still, as if capable of something, I stared again at lost bile beneath a salt-rimed car—seemingly motionless but rotting steel: wild molecular alteration—and worried for the earth's simple, ungravelled glaciers and for the Arctic, for the scarred tundra above the DEW Line (where I'd spent that year scanning USAF radar for Russian bombers), for the sweet moss there, all the water, the unexpected birds, my racing fox whose fiery coat had skipped (for me?) an entire change of season.

With a long look at the waxed sky, I moved to the house. Inside, I arranged Alex's and my boots on the doormat. The expensive heated air in my house flushed my face.

—DONALD ANDERSON, "Fire Road"

Of the six surviving subspecies of tiger, the Amur is the only one habituated to arctic conditions. In addition to having a larger skull than the other subspecies, it carries more fat and a heavier coat, and these give it a rugged, primitive burliness that is missing from its sleeker tropical cousins. The thickly maned head can be as broad as a man's chest and shoulders, and winter paw prints are described using hats and pot lids for comparison. As the encyclopedic reference *Mammals of the Soviet Union* puts it, "The general appearance of the tiger is that of a huge physical force and quiet confidence, combined with a rather heavy grace." But one could just as easily say: this is what you get when you pair the agility and appetites of a cat with the mass of an industrial refrigerator.

—JOHN VAILLANT, *The Tiger: A True Story of Vengeance and Survival*

Winter has no entrails. At its heart is an echo.

—EDWARD LUEDERS, *The Clam Lake Papers:
A Winter in the North Woods*

By now the tiger had not rested or eaten well in a week. This would not have been quite so serious had it been a different season, but the temperature was ranging from twenty-five to forty-five below zero. The amount of meat required to keep something the size of a tiger as much as 150 degrees hotter than the world around was prodigious—on the order of forty pounds per day. . . . The impact of an attacking tiger can be compared to that of a piano falling on you from a second-story window. But unlike the piano, the tiger is designed to do this, and the impact is only the beginning.

—JOHN VAILLANT, *The Tiger: A True Story of Vengeance and Survival*

There are, of course, *four* Poles, two set—*in place*—North and South, geographically, and two, North and South, magnetically, that *move*, affected, always, by the Earth's own wanderings.

Then came March 24, 1989. The 987-foot supertanker *Exxon Valdez* ran aground on Bligh Reef with 1,547,180 barrels of crude on board. The captain, almost certainly drunk, had been trying to avoid icebergs that had calved off the Columbia Glacier. Almost eleven million gallons of crude ran into Prince William Sound. Remember the television pictures of oil-stricken sea otters and tar-strangulated gulls? But the Exxon PR machine was ready. The Russian solution to accidents was to nuke the evidence. The American way was to spin the message. But the net result was the same: environmental disaster. Exxon beamed images around the world of flotillas of omni-sweepers and maxi-barges sucking up spilled oil, and of fifteen thousand workers scrubbing rocks. (After the few days PR gurus decided the scrubbers weren't cleaning, they were treating.) One could imagine the oilmen Laurel and Hardy joining in, Hardy scrubbing industriously and Laurel subverting his efforts. Exxon released footage of 250 sea otters being flown to rehabilitation centers where they dined on crab; to the joy of the world, all 250 survived. It cost Exxon $90,000 per animal. The rest of the otters died. If the PR men were smirking, the joke was on us. Private contractors working on the cleanup became known as spillionaires.

—SARA WHEELER, *Magnetic North: Notes from the Arctic Circle*

Undoubtedly the polar bears at the zoo both dream that all other animals will discover, upon waking, their bodies buried in snow.

—JYNNE DILLING MARTIN, from
"Reasons to Consider Setting Ourselves on Fire"

If you want to photograph a polar bear sleeping on an ice floe, you will need very good eyes. Your heat sensitive camera won't help you find one because the variable blubber layer that protects polar bears from heat loss works so well that they "disappear" from the heat sensitive radar. The naked eye might not work that well either: their fur is not actually white; it is transparent, made of hollow individual hairs that reflect as white to the human eye. These hollow hairs carry the sun's short-wave-length energy to the bear's black skin, an element in the polar bear's unique and sophisticated heat-exchange and management system that can keep the skin 15 degrees warmer than its surround-ings. This complex and intriguing system has evolved specifically for living on the ice, and for constant movement in and out of frigid waters.

—PAULINE COUTURE, Ice

Into the eternal darkness, into fire and into ice.

—DANTE ALIGHIERI, The Divine Comedy, "Inferno," Canto III

PART II

A man said to the universe:
"Sir I exist!"
"However," replied the universe,
"The fact has not created in me
A sense of obligation."

—STEPHEN CRANE, from *War is Kind*

In the end, the trip was mostly a calculated and well-crafted presidential publicity stunt. And it raised the question: If the American people see the President of the United States standing atop a melting glacier and telling them the world is in trouble, will they care?

—JEFF GOODELL, from *The Rolling Stone Interview* with Obama,

October 8, 2015

A *yaranga* is a tentlike traditional mobile home of nomads of some Northern indigenous peoples of Russia, such as Chukchi. Built of a light wooden frame (historically whale bones) covered with reindeer skins sewn together, a medium-size *yaranga* requires about fifty skins. *Yaranga* originates from the Persian word *yarangah*. During the height of the Persian Empire, Persian was spoken from the Arabian Peninsula to Siberia. *Yaran* means "companion/guest" while *gah* means "a place."

Reindeer hair insulates so well that the blood and organs of a dead deer will not freeze under intact skin. The blood and organs will, instead, ferment. Polar bear hair is also hollow.

Down is the soft under plumage (a layer of insulation underneath feathers) that keeps waterfowl warm and dry. Unlike feathers, down is un-quilled, and lends itself for "fill" for clothing and covers.

The secret of GORE-TEX® products—which are both completely water-proof and completely breathable at the same time—lies within its revolution-ary bi-component membrane. The membrane contains over nine billion microscopic pores which are approximately 20,000 times smaller than a drop of water, but 700 times bigger than a molecule of moisture vapor. So while water in its liquid form cannot penetrate the GORE-TEX® membrane, as moisture vapor it can easily escape.

—see http://www.gore-tex.com

The GORE-TEX® membrane is a polymer, as are, in fact, wool and silk, though un-synthetically.

> Himalayan glaciers, seen as water towers for Asia, have been shrinking at a rate of 0.5 percent a year—similar to glaciers in South America's Andes and the European Alps. The Himalayan shrinkage is being studied by the University of Colorado's Institute of Arctic and Alpine Research. Oxygen in snow gains mass as snow becomes glacier ice. The changes are reflected in different numbers of neutrons, which become signatures used to identify glacier water in rivers. Among the rivers starting in the Himalayas, the Indus, Amu Darya, Brahmaputra, Ganges, Mekong, Yangtze, and Yellow, are, of course, critical sources for people in Afghanistan, Bangladesh, China, India, Nepal, and Pakistan. . . .
> —see BRUCE FINLEY, *The Denver Post*, December 18, 2011

> A Yup´ik hunter on Saint Lawrence Island once told me that what traditional Eskimos fear most about us is the extent of our power to alter the land, the scale of that power. . . . They call us, with a mixture of incredulity and apprehension, "the people who change nature."
> —BARRY LOPEZ, *Arctic Dreams*

Is it common knowledge that, prior to organ harvesting, the harvester will inject ice-cold water into the donor's ear to confirm that the donor is in fact, contrary to the other option, dead?

> The snow came down last night like moths
> Burned on the moon; it fell till dawn,
> Covered the town with simple cloths.
>
> Absolute snow lies rumpled on
> What shellbursts scattered and deranged,
> Entangled railings, crevassed lawn.
>
> As if it did not know they'd changed,
> Snow smoothly clasps the roofs of homes
> Fear-gutted, trustless and estranged.

The ration stacks are milky domes;
Across the ammunition pile
The snow has climbed in sparkling combs.

You think: beyond the town a mile
Or two, this snowfall fills the eyes
Of soldiers dead a little while.

Persons and persons in disguise,
Walking the new air white and fine,
Trade glances quick with shared surprise.

At children's windows, heaped, benign,
As always, winter shines the most,
And frost makes marvelous designs.

The night guard coming from his post,
Ten first-snows back in thought, walks slow
And warms him with a boyish boast:

He was the first to see the snow.
                    —RICHARD WILBUR, "First Snow in Alsace"

In far-back mountain coves they say that walking barefoot in the
snow, the first snow of the season, will prevent you catching cold all
year. Ten minutes stepping through the woods will do the work, as
though the shock of bare soles on the ice will make the flesh immune
for months. Perhaps the surge of energy to fight the freezing wetness
spurs resistance, or maybe the humility of going pilgrim-like into the
storm while others hide from frost and bitter elements inspires the
blood to militance, to strength. The naked soles in snow may be like
feet that step on beds of coals in mystical ceremonies to show that
mind and nerves are free above the pain's immensity.
                    —ROBERT MORGAN, "Immune"

The wind had died. The snow was falling straight down, less of it now, and lighter. We drove away from the resort, right up to the barricade. "Move it," my father told me. When I looked at him he said, "What are you waiting for?" I got out and dragged one of the sawhorses aside, then put it back after he drove through. He pushed the door open for me. "Now you're an accomplice," he said. "We go down together." He put the car in gear and gave me a look. "Joke, son."

Down the first long stretch I watched the road behind us, to see if the trooper was on our tail. The barricade vanished. Then there was nothing but snow: snow on the road, snow kicking up from the chains, snow on the trees, snow in the sky; and our trail in the snow. Then I faced forward and had a shock. The lay of the road behind us had been marked by our own tracks, but there were no tracks ahead of us. My father was breaking virgin snow between a line of tall trees. He was humming "Stars Fell on Alabama." I felt snow brush along the floorboards under my feet. To keep my hands from shaking, I clamped them between my knees.

My father grunted in a thoughtful way and said, "Don't ever try this yourself."

—TOBIAS WOLFF, "Powder"

April 2012. An avalanche that buried more than 120 Pakistani soldiers in a Himalayan region close to India has spotlighted what critics refer to as one of the world's most pointless military deployments: two poverty-wracked nations engaged in a costly standoff over an uninhabitable patch of mountain and ice. Since the massive wall of snow engulfed a Pakistani military complex close to the Siachen Glacier, rescue teams have been unable to dig up any of the buried troops. The conflict over Siachen began in 1984 when India occupied the heights of the 49-mile long glacier, fearing Pakistan wanted to claim the territory. Islamabad also deployed its troops. A 2003 cease-fire largely ended skirmishes on the glacier, where troops have been deployed as high as 20,000 feet, but both armies have remained encamped. Temperatures as low as minus 76°F, vicious winds and altitude sickness (the region is just east of the world's second-highest peak, K-2) have killed far more than the artillery fire. The avalanche plowed into the Pakistani headquarters at the entrance to the glacier, and buried the

complex under more than 70 feet of snow. Publicly, the army has held
out hope of survivors.

—"US Military Experts to Assist Pakistan Avalanche Rescue Efforts,"
Fox News, April 4, 2012, http://www.foxnews.com/world/2012/04/10/
us-military-experts-to-assist-pakistan-avalanche-rescue-efforts.html

I thought ahead, and that was why I knew there would be other troop-
ers waiting for us at the end of our ride, if we even got there. What I
did not know was that my father would wheedle and plead his way
past them—he didn't sing "O Tannenbaum," but just about—and
get me home for dinner, buying a little more time before my mother
decided to make the split final. I knew we'd get caught; I was resigned
to it. And maybe for this reason I stopped moping and began to enjoy
myself.

Why not? This was one for the books. Like being in a speed-
boat, only better. You can't go downhill in a boat. And it was all ours.
And it kept coming, the laden trees, the unbroken surface of snow,
the sudden white vistas. Here and there I saw hints of the road,
ditches, fences, stakes, but not so many that I could have found my
way. But then I didn't have to. My father in his forty-eighth year,
rumpled, kind, bankrupt of honor, flushed with certainty. He was a
great driver. All persuasion, no coercion. Such subtlety at the wheel,
such tactful pedalwork. I actually trusted him. And the best was yet
to come—switchbacks and hairpins impossible to describe. Except
maybe to say this: if you haven't driven fresh powder, you haven't
driven.

—TOBIAS WOLFF, "Powder"

They were sitting in bed. New Hampshire farmers were preparing
snowplows in barns and garages all down the highway and across the
hills. The thermometer was stuck at zero. And it seemed unusually
dark. Hunt told Leah that he thought he could almost hear the ice
freezing on the lake outside, hardening with rigid snaps. Leah said
that what he heard was the start of deer season. Then some time near
dawn, Hunt told her. Leah shook. He tried to hold her. She fought
him. Fought and then got up to pack.

"I'm going to take the children," she said.

"Where?"

"To tell your parents."

"It's still dark."

"I don't care."

"I do."

"You've made your choice."

But Hunt held Leah back until after breakfast. Snow had started. The sky looked like soft, faded stone. They put one bag and the children in the back of the wagon and set off for Boston to tell his parents about divorce. Every fourth car heading north, passing or passed by them, was racked and spread with a dead deer. "Isn't it awful!" one of the children said, as they passed a camper laden with two. In the rearview mirror, Hunt could see blood running down over the camper's windshield.

—DAVID KRANES, "Hunt"

Frederic Edwin Church, in one of his most heroic attempts to portray transcendent, inutile grandeur, rendered with his painstaking brilliance the icebergs of the North Atlantic. Now, with Greenland crisscrossed by commercial air routes and the Himalayas littered with empty oxygen canisters, Antarctica is the Sublime's last stronghold, where Man can still be cowed by the inhuman.

—JOHN UPDIKE, from Still Looking: Essays in American Art

Entr'acte

A three-week-old blizzard, become manageable in our yards, remained intact among the pine and scrub oak. The tree-bound drifts had disposed themselves to predictable depths. Crusted by winds and no sun, the snow (in places two and three feet deep) had supported Alex, and I paralleled her steps, cracking through the shell which had afforded my daughter passage. Though I'd dug out another hat and gloves, and had worn my coat, I'd not worn boots. I'd laced up low-quartered, nylon running shoes.

The snow beneath the crust was soft and sucked at my feet and frightened me with its pull. I felt lost as a child. I began to run uphill, but kept crashing through the crust, banging my shins, into my own holes.

Intersecting the lower fire road, I squatted, blowing, in a beaten trail, snow-packed by the passage of the four-wheelers, x-country skiers, winter joggers, sleds. But, night fallen, I had become the single human presence. The trees behind blocked all the houses, and the rising ridge still concealed the mall. For all the civilization I knew existed, I felt alone on what could have been, except for the trees, the Polar Ice Cap.

Though I could not see it, I sensed the rise of the moon. I longed for the arrival of my daughter, striding through the dark to end my search, to lead and take me home. I removed my gloves, stood and shook my trousers. Then, knowing more than feeling, hitched my pants to expose my chipped shins and recalled the news report of the Hungarian religious fanatic who in madness, screaming, I AM JESUS CHRIST!, scaled a marble balustrade to attack Michelangelo's Pietà with a car mechanic's hammer, cleaving the Madonna's nose, break-ing her left arm (milk-white marble fingers snapping as the left hand struck the floor) in a Vatican basilica. But the image of the broken Madonna was wrong, the comparison vain: (1) I was no damaged mother; (2) had been impregnated by no god; and (3) my offspring, like anyone's, imperfect. All the same, Mary, too, it sunk in, was a sin-gle parent, bedazed Joseph a pathetic stand-in. . . . I quit to examine my legs. My foolish thrashing in crusted snow had affected no one, not even me. The shins would heal.

—DONALD ANDERSON, "Fire Road"

The first icebergs we had seen, just north of the Strait of Belle Isle, listing and guttered by the ocean, seemed immensely sad, exhausted by some unknown calamity. We sailed past them. Farther north they began to seem like stragglers fallen behind an army, drifting, self-absorbed, in the water, bleak and immense. It was as if they had been borne down from a world of myth, some Götterdämmerung of noise and catastrophe. Fallen pieces of the moon.

—BARRY LOPEZ, *Arctic Dreams*

The Antarctic ice cap is the largest single chunk of ice on earth, containing mountains that are two thousand miles in length, with icy winds that can blow in excess of two hundred miles per hour. To this day, there are no permanent human settlements on Antarctica. As spring arrives, Antarctica sees the Adelie penguins arrive from the sea, ready to breed. But as the spring arrives, so do the katabatic winds, with blinding snow and ice so severe that many of the nesting penguins freeze to death in place. Other than the katabatic winds, there are no natural "land" predators in Antarctica to threaten the penguin population.

We carry on as if we live at the end of time, but as [the Greenland ice] cores reveal, we actually exist in a continuum and are less important than we care to think. This awareness [has] survived among the Inuit, people who live peacefully with the certainty of their own cosmic insignificance—not something that could be said of Robert Peary.

—SARA WHEELER, *Magnetic North: Notes from the Arctic Circle*

The neurotic and megalomaniacal Peary had staying power, at least: on expedition after expedition he smashed his way through Smith Sound to the tip of Greenland, marching on even after eight of his toes snapped off. On April 6, 1909, he and his black manservant Matthew Henson stood on the ice at 90°N—or so Peary claimed. It was, he wrote in his journal, "the finish, the cap and climax, of nearly 400 years of effort, loss of life, and expenditure of fortunes by the civilized nations of the world, and it has been accomplished in a way that is thoroughly American." Both Peary and Henson left offspring in the Arctic, taking away instead three thirty-seven-ton meteorites sacred to the Greenlanders, as well as six live Inuit specimens, four of whom quickly picked up alien germs and died in Washington. The meteorites are in the Smithsonian. One of the surviving Inuit, the boy Minik, found out that his father's bones were on display at the Natural History Museum in New York.

—Ibid.

Contemporary polar historians agree that Peary, despite his avowals and lost toes, did *not* reach the Pole. Nonetheless, his mythology as one of America's best known explorers has persisted, outliving his lousy personality and fake achievements.

One estimate suggests that, at minus 40 degrees Fahrenheit, half of what someone eats is utilized by the body just to maintain its proper core temperature. According to another study, at minus 30 degrees Fahrenheit, men expend up to one thousand calories simply to warm and humidify the air they breathe. When the men of the *Discovery* expedition felt incipient starvation far too quickly in their first summer of manhauling, Scott upped their food allowance from 28.6 ounces to 35.5 ounces, with some success. That ration gave 3,971 calories per day. It's thought, however, that an average of 6,500 calories would be necessary for the hardest days of Antarctic manhauling, an ordeal estimated to require more energy per day than competing in the Tour de France.

—JASON C. ANTHONY, *HOOSH: Roast Penguin, Scurvy Day, and Other Stories of Antarctic Cuisine*

Before mechanical refrigeration systems were introduced, people cooled their food with ice and snow, either found locally or brought down from the mountains. The first cellars were holes dug into the ground and lined with wood or straw and packed with snow and ice: this was the only means of refrigeration for most of history. Refrigeration is the process of removing heat from an enclosed space, or from a substance, to lower its temperature. A refrigerator uses the evaporation of a liquid to absorb heat. The liquid, or refrigerant, used in a refrigerator evaporates at an extremely low temperature, creating freezing temperatures inside the refrigerator. It's all based on the following physics:—a liquid is rapidly vaporized (through compression)—the quickly expanding vapor requires kinetic energy and draws the energy needed from the immediate area—which loses energy and becomes cooler.

—"The History of the Refrigerator and Freezers," ThoughtCo.com, http://inventors.about.com/library/inventors/blrefrigerator.htm

I date the end of the old republic and the birth of the empire to the invention, in the late thirties, of air conditioning. Before air conditioning, Washington was deserted from mid-June to September. . . . But after air conditioning and the Second World War arrived, more or less at the same time, Congress sits and sits while the presidents or at least their staffs never stop making mischief.

—GORE VIDAL, *The Essential Gore Vidal*

The Carrier Dome, situated at Syracuse University, is the largest domed stadium of any college campus. The Dome was financed by the Carrier Corporation, the brainchild of Willis Haviland Carrier, who is correctly credited with inventing modern air conditioning in 1902. Carrier moved his company's headquarters to Syracuse, New York in 1930. The term "air conditioning" came later. Carrier labeled his first patent "Apparatus for Treating Air."

Two thousand miles from the slowly moving wagons of the Donner Party, a storm rose from the unruly waters of the Gulf of Alaska. Evaporation sent countless tons of water vapor pirouetting up into the sky, loading up the clouds like soaked sponges. Winds gathered strength until they howled across the whitecaps like a flood down a gully. The jet stream began to push the front southward, shoving it along at the speed of a hardworking freight train.

The storm smashed into the redwood forests at the mouth of the Humboldt River, almost up by Oregon, dropping sheets of rain that soaked deep down into the roots, ensuring another lush green winter at the ocean's edge. The heart of the maelstrom rushed farther south, cascading over the coast ranges, whipping through the Golden Gate, casting long and threatening shadows across Sutter's Fort. At last the clouds began their long climb up into the Sierra, aiming directly for the pass the emigrants intended to use. As the clouds rose, they cooled. Gaseous molecules condensed into liquid droplets; droplets froze into ice crystals; crystals coalesced into snowflakes. The flakes grew too heavy to resist the call of gravity. It began to snow.

—ETHAN RARICK, *Desperate Passage:*
The Donner Party's Perilous Journey West

The earliest meteorological records for the Sierra owe their existence not to a desire to measure the range's ferocity but to conquer it. When the transcontinental railroad subdued the Sierra with ties and track, Southern Pacific maintenance crews began to record weather conditions on a daily basis. Fortunately for the study of the Donner Party, the line goes right over the point at which the emigrants aimed to cross the ridge of the Sierra. Thus we have weather records for the spot dating back to 1870, the longest stream of meteorological data for any location in the Sierra.

The resulting picture is one of almost mythical snowfall. In 1938, apparently using a particularly conservative method of measurement, the railroad crews recorded a winter-long total of sixty-eight feet. Because snow compacts—the heavier new snow on top pressing down the old layer beneath—the amount on the ground at any one time is less than the seasonal totals. It is not uncommon to have fifteen feet of snow on the level and that is far from the maximum. In 1880, just a decade after the railroad opened, the diligent observers of the Southern Pacific recorded a depth at Donner Pass of thirty-one feet, enough to cover a three-story building.

—Ibid.

That night, having made six miles, they camped at the brink of the canyon. No celebration brought in the New Year. In the morning they fell rather than walked to the bottom. On slopes which were not too dangerous they squatted on their snow-shoes and slid, but on coming to the end of the slope they usually plunged deeply into a drift. Then encumbered with pack and shoes, and faint with weakness, they had to fight their way to their feet. Getting to the bottom did not take long, and they found the stream low because of the freezing of its headwaters. They crossed without difficulty, probably upon a snow-bridge. But going up the other side was a different tale. For fifty feet up from the river the wall was precipitous; they climbed up by holding on to crevices and bushes. Then came a slope where trees could just cling. Digging their snow-shoes into the surface they had to work their way up as if ascending a stairway, each step a struggle. Blood from their feet marked the white snow. Completely exhausted after one of the hardest days they had yet endured, they just managed to get to the top of the canyon by evening. Fosdick was the weakest, and could scarcely make it. That night they ate the last of the dried human flesh which, loathsome as it was, had been their only food for a week.

They found themselves next morning upon a sort of plateau. The country was less rugged; the sun shone more warmly; and the surface of the snow was so firmly consolidated that they found themselves able to walk without snow-shoes. But to offset these advantages they were again without food, and Fosdick held them back. Their feet were worse than ever, and the toes of one of the Indian boys began drop-ping off at the first joints. The snow where they camped was only six feet deep.

—GEORGE R. STEWART, *Ordeal by Hunger:*
The Story of the Donner Party

The first warm day,
and by mid-afternoon
the snow is no more
than a washing
strewn over the yards,
the bedding rolled in knots
and leaking water,
the white shirts lying
under the evergreens.
Through the heaviest drifts
rise autumn's fallen
bicycles, small carnivals
of paint and chrome,
the Octopus
and Tilt-A-Whirl
beginning to turn
in the sun. Now children,
stiffened by winter
and dressed, somehow,
like old men, mutter
and bend to the work
of building dams.

But such a spring is brief;
by five o'clock
the chill of sundown,
darkness, the blue TVs
flashing like storms
in the picture windows,
the yards gone gray,
the wet dogs barking
at nothing. Far off
across the cornfields
staked for streets and sewers,
the body of a farmer
missing since fall
will show up
in his garden tomorrow,
as unexpected
as a tulip.

—TED KOOSER, "Late February"

In the early morning the wind blew sheets of snow against the tent. Mawson ached for food but first answered the constant nag on his mind—the weight of the sledge. The bones would be boiled down to marrow jelly, and then "the skeleton could be safely jettisoned." After simmering out the skimpy nutriment from Ginger's depleted marrow, one source of food was left that he could not ignore—Ginger's head.

It made a scene that burned into his memory. Two haggard, tattered men, crouched in the narrow tent, watching the skinned dog's head cooking.

When Ginger's head was boiled for ninety minutes, they lifted it with the two wooden spoons onto the lid of the cooker. Mawson ran a knife across the top of the skull for the demarcation line. They took turns in gnawing their different sides, biting away the jaw muscles and lips, swallowing the eyelids and gulping down the eyeballs. With the wooden spoons they scooped out the contents of the skull—and then split the tongue, the thyroid, and the brain into two servings; only their fierce hunger made them grateful for this macabre feast, only the pangs of starvation made it possible for Mawson to write before sleeping: "Had a good breakfast from Ginger's skull—ate brains, thyroid, and all."

—LENNARD BICKEL, *Mawson's Will:*
The Greatest Polar Survival Story Ever Written

By 1880, refrigeration was cheaper than natural ice. With efficient refrigeration, food could be shipped to markets. Hog production grew. Over a few years, beef exports from the United States to the British Isles grew from a hundred thousand pounds a year to seventy-two million pounds a year. More than a hundred thousand refrigerated railroad cars appeared almost overnight. Suddenly, midwestern farmers could undercut New England farmers on dairy products. Refrigeration increased the profitability of ranching, and ranches expanded, implicating refrigeration in the last phase of the eviction of Native Americans from their ancestral lands and the near extinction of the buffalo. Clarence Birdseye, inspired by the quality of fish frozen at forty below after ice fishing, developed techniques for flash freezing food in 1923. By 1928, Americans were buying a million pounds of flash-frozen food a year. With air-conditioning, people could live comfortably in Florida and Arizona and New Mexico. By virtue of the air conditioner, skyscrapers grew taller, into the range of gusting winds that prevented the opening of windows.

—BILL STREEVER, *Cold*

Immediately I recognized the huge, sprawling bulk of Kanchenjunga, at 28,169 feet above sea level the third-highest mountain on earth. Fifteen minutes later, Makalu, the world's fifth-highest peak, came into view—and then, finally, the unmistakable profile of Everest itself.

The ink-black wedge of the summit pyramid stood out in stark relief, towering over the surrounding ridges. Thrust high into the jet stream, the mountain ripped a visible gash in the 120-knot hurricane, sending forth a plume of ice crystals that trailed to the east like a long silk scarf. As I gazed across the sky at this contrail, it occurred to me that the top of Everest was precisely the same height as the pressurized jet bearing me through the heavens. That I proposed to climb to the cruising altitude of an Airbus 300 jetliner struck me, at that moment, as preposterous, or worse. My palms felt clammy.

—JON KRAKAUER, *Into Thin Air:*
A Personal Account of the Mount Everest Disaster

And then there is the space shuttle *Challenger*. Hot gas leaked around cold O-rings, causing an explosion of liquid hydrogen and liquid oxygen and killing seven astronauts seventy-three seconds after liftoff. The temperature on the launchpad that January morning in 1986 was thirty-six degrees, fifteen degrees colder than the conditions under which the rocket's O-rings were tested.

—BILL STREEVER, *Cold*

People in Florida hate the cold. Many have moved here from places such as Detroit and Minneapolis and Montana. My cabdriver, originally from Chicago, brags that he no longer owns a warm coat. Near the hotel pool, a woman who grew up in Florida tells me that she cannot even imagine forty below.

"Imagine thirty degrees," I tell her. "Now imagine eighty degrees. They are fifty degrees apart. Now imagine that same difference, but downward from thirty degrees. That is twenty below. Subtract another twenty degrees, and you have forty below." She stares blankly, as if at a madman. Imagination cannot extrapolate beyond the temperatures it has experienced. I stand in the sun, face upward, enjoying for the moment a latitudinal spring.

—Ibid.

Mount Washington is 6,288 feet high, and the summit and its buttress ridges lie at the intersection of two jet streams in the upper air. These combine with the upslope on the windward side of the range and the high elevation of the crests to create local weather systems of sudden and extraordinary violence. In later years, scientists would occupy the summit of Mount Washington and measure temperatures of 60° below zero and winds of 231 mph, the highest ever recorded on the surface of the earth.

—NICHOLAS HOWE, *Not Without Peril*

The observatory's 24-hour data sheet shows that the summit was in the clouds with a midnight temperature of 25 degrees and a southeast wind of 99 miles per hour, and rime ice was forming on the windward side of every outdoor surface. At 3 A. M. the temperature had gone up one degree and the wind had reached 120 mph. Thirty-five minutes later it hit 160.

I was still awake. The observatory had a good lounge and I'd been reading a book by one of the 19th-century observers who told of a day of high wind when the crew was wondering if their building would be blown apart. As a precaution, one of them wrapped himself in a mattress and stiffened the package by attaching crowbars to it.

The violent weather did not worry me. While the tempest outside was tearing at the building, I was sitting on an upholstered sofa in the well-heated lounge and watching television. . . . I didn't think I was in danger, because the walls of the observatory had withstood every previous storm and the thick plate glass windows were protected by heavy steel grating. . . . Ice was forming too fast for the deicers that were protecting a set of instruments at the top of the observatory's tower. Bill went up the inside of the tower to knock off the worst of the ice. He'd take a full swing with a crowbar, but sometimes he would miss the ice. The wind was that strong even in the lulls.

I went up with him, wearing an experimental facemask that the observatory was testing. It was like a military gas mask except there was a round opening in front that was held by a latch and could be opened for eating or spitting. The makers had not anticipated weather like this. When I opened the latch and faced the wind it seemed as if the air rushing in would inflate me; when I turned around, the wind seemed to be pulling away so fast I wasn't sure I could draw a breath from it. I seemed to be drowning in an ocean of air.

—Ibid.

All of us have dark stirrings of doubt and fear whenever the Donner Party is mentioned. In such extremis what would we do? Snow-trapped and starving in the Sierras with no hope of relief, would we fall to devouring each other . . . In the purely physical realm of survival, what justifies what?

—JAMES DICKEY, qtd. in Ethan Rarick, *Desperate Passage: The Donner Party's Perilous Journey West*

The last Mount Washington death of the 1994 winter was in June. As the sun climbs toward summer it loosens the ice that forms on the ledges lining the ravine at the level of Schiller's Rock. It's on just such lovely days that the greatest number of skiers come up, and long practice has endowed a citizen's early warning system: when the telltale crack is heard, the cry "ICE!" goes up. Sound carries well in this vast acoustic focus, everyone hears the call, and everyone looks up the slope.

On June 4, Sarah Nicholson looked up and saw a car-sized block of ice sliding and bounding down toward her. Gravity is also on the side of the skier, and a quick escape left or right downslope almost always avoids the danger of falling ice. But this block was breaking into fragments, and it wasn't clear which way led most quickly to safety. It's a familiar sidewalk dilemma: step left or right to avoid the collision? Sarah's moment of hesitation broke the heart of her friends, and brought the list of mortality on the Presidential Range to 115.

—NICHOLAS HOWE, *Not Without Peril*

They had, from necessity, come to eat almost every part of the body. Canessa knew that the liver contained the reserve of vitamins; for that reason he ate it himself and encouraged others to do so until it was set aside for the expeditionaries. Having overcome their revulsion against eating the liver, it was easier to move on to the heart, kidneys, and intestines. It was less extraordinary for them to do this than it might have been for a European or a North American, because it was common in Uruguay to eat the intestines and the lymphatic glands of a steer at an asado. The sheets of fat which had been cut from the body were dried in the sun until a crust formed, and then they were eaten by everyone. It was a source of energy and, though not as popular as the meat, was outside the rationing, as were the odd bits of earlier carcasses which had been left around in the snow and could be scavenged by anyone. This helped fill the stomachs of those who were hungry, for it was only the expeditionaries who ever ate their fill of meat. The others felt a continuous craving for more, yet realized how important it was that what they had should be rationed. Only the lungs, the skin, the head, and the genitals of the corpses were thrown aside.

—PIERS PAUL READ, *Alive: The Story of the Andes Survivors*

I arrived on the South Col, our launching pad for the summit assault, at 1:00 P. M. A forlorn plateau of bulletproof ice and windswept boulders 26,000 feet above sea level, it occupies a broad notch between the upper ramparts of Lhotse and Everest. Roughly rectangular in shape, about four football fields long by two across, the Col's eastern margin drops 7,000 feet down the Kangshung Face into Tibet; the other side plunges 4,000 feet to the Western Cwm. Just back from the lip of this chasm, at the Col's westernmost edge, the tents of Camp Four squatted on a patch of barren ground surrounded by more than a thousand discarded oxygen canisters. If there is a more desolate, inhospitable habitation anywhere on the planet, I hope never to see it.

—JON KRAKAUER, *Into Thin Air:*
A Personal Account of the Mount Everest Disaster

Marriage is not
a house or even a tent

it is before that, and colder:

The edge of the forest, the edge
of the desert
                    the unpainted stairs
at the back where we squat
outside, eating popcorn

the edge of the receding glacier

where painfully and with wonder
at having survived even
this far

we are learning to make fire.

—MARGARET ATWOOD, "Habitation"

Entr'acte

Bare-handed, I crushed white snow against the wounds, the snow blackening, not altogether unpleasantly, in my hands. But that done, the darkness about me swelled, and my hands went cold, and hunkering alone in the frozen trail, I would have settled for any sign of another's life other than the trampled snow I bent in: a trace of granola, a butt, an oil leak, the visible track of a bird.

I pictured ravens, then redwing blackbirds, large as crows. I pictured white ponds. Sloughs. Frozen, ratty cattails. When a blackbird attempted to perch, the cattail snapped. Near frozen inlets, I pictured turtles, on their backs. I stopped myself, took hold, inhaled, rubbed the backs of my hands on my nose. I forced deep breaths, felt things catch in my chest, slow. I wished that I hadn't been drinking.

As a child, snowshoeing in a storm, I had fallen behind my father, and had fallen. The snow gave to the curve of my back and my beetle weight and I'd bellowed for my father. He tramped back, stooped, grasped my wrists. "Stop flailing. You look foolish," he said. "Like a bug." He held onto my wrists, not to raise me, but to remove my gloves. He shook snow from them. Having bared my hands, he instructed me to choose and unhitch one snowshoe which I was to use to support my weight as I rolled with it pressed to my chest. It worked, and I stood.

"Now rehitch it."

But my hands were cold and the leather thongs of the snowshoe were frozen. My father ungloved his hands and rehitched the shoe. Then, as if I'd not fallen and I'd not cried and he'd not bent to assist me, my father turned and broke trail.

A father myself (and older by years than mine at the time we'd hiked that snow), I rose and pursued the fire road, knowing it to soon turn and rise in its final, short push for the low summit of our civilized wild ridge. Then there Alex was: Alex and a single bag of food—the size of a year-old child—which she clutched to her chest. She walked, staring down. My heart jumped. Seeing her touched as deep a feeling as I'd ever known gaping at her. The day she'd been born, I'd gaped. Our first time together had been without the mother too—just Alex

(weightless in my arms) and me in a corridor in a polished hospital at
dawn, in Utah. Then, too, there had been a full moon.

—DONALD ANDERSON, "Fire Road"

At the same time as the boys dug into the snow in search of the buried
bodies, the corpses that they had preserved nearer the surface began
to suffer from the stronger sun which melted the thin layer of snow
which covered them. The thaw had truly set in—the level of the snow
had fallen far below the roof of the Fairchild—and the sun in the middle of the day became so hot that any meat left exposed to it would
quickly rot. Added, then, to the labors of digging, cutting, and snow
melting was that of covering the bodies with snow and then shielding
them from the sun with sheets of cardboard and plastic.

As the supplies grew short, an order went out from the cousins
that there was to be no more pilfering. This edict was no more effective than most others which seek to upset an established practice. They
therefore sought to make what food they had last longer by eating
parts of the human body which previously they had left aside. The
hands and feet, for example, had flesh beneath the skin which could be
scraped off the bone. They tried, too, to eat the tongue off one corpse
but could not swallow it, and one of them once ate the testicles.

On the other hand they all took to the marrow. When the last
shred of meat had been scraped off a bone it would be cracked open
with the ax and the marrow extracted with a piece of wire or a knife
and shared. They also ate the blood clots which they found around the
hearts of almost all the bodies. Their texture and taste were different
from that of the flesh or fat, and by now they were sick to death of this
staple diet.

—PIERS PAUL READ, Alive: The Story of the Andes Survivors

Lhotse Face: What the western flank of Lhotse is termed. Any climber
bound for the South Col on Everest must climb this 3,700 foot wall of glacial
blue ice. The face rises at 40- and 50-degree pitches with the occasional
80-degree bulges.

Temperatures on Everest can fall below -100°F with winds gusting to the strength of a category 3 hurricane (118 miles per hour or more). As you progress up Everest, the amount of oxygen available to sustain life decreases until you reach the summit, where there is only a third of the level of oxygen required to sustain human life. Enroute, most climbers tend to agree that the hardest part of the mountain to climb is the Khumbu Icefall, just above base camp. The icefall is also one of the most treacherous stages of the South Col route. The Khumbu Icefall is the point in the glacier where it begins to melt. Because of this, it is like trying to climb over a moving sea of ice. The entire glacier is immensely unstable, and the average climber can expect to be in this danger zone for between four and six hours before they ascend 2,000 feet to the top of the glacier where Camp One is located. The Khumbu Icefall requires clipping and unclipping from fixed lines (ropes attached to the mountain with pickets buried in the snow), and crossing dark, bottomless crevasses on nothing more than a regular old garden ladder, all while wearing crampons on your feet. With the glacier moving constantly, crevasses open with no warning and blocks of ice, up to the size of a house, can fall from above. Blocks of ice shift; large pieces can crack off and tumble down through the icefall forming a maze of pieces that the climber must work quickly to navigate. The Khumbu Icefall is responsible for some ten percent of the 200 known deaths on the mountain.

— "The Khumbu Icefall: Mount Everest's Death Trap," Scrilbol.com, http://scribol.com/environment/lakes-and-rivers/the-khumbu-icefall-mount-everests-death-trap/

The Antarctic is alien land. Desolate and barren, hostile to life, it is a lost continent at the bottom of our world, now smothered under the greatest ice shield known.

This solid ice cap is an immensity. Three miles thick in places, with a mean, overall thickness of a mile and a quarter, it blankets a region bigger than Europe and the United States combined—almost six million square miles of the earth's surface in southern summer. In winter, when the sea freezes over, the area inundated with ice can be doubled. On the mainland, from the high and remote solitude of the South Pole and the so-called Pole of Inaccessibility down to the coastal fringes, only two percent of the rocky land is able to break free of the frozen mantle.

Far inland, the ice plains rise more than 12,000 feet above sea level, and only the frost-rimed peaks of the mightiest mountains can pierce this frigid shield. Cloaking, submerging the great ranges, the ice makes the Antarctic the highest overall continent on earth and exerts an influence and an impact on the world's weather to an extent not yet fully understood.

—LENNARD BICKEL, *Mawson's Will:*
The Greatest Polar Survival Story Ever Written

"Both bodies were partially buried," Hutchison recalls. "Their backpacks were maybe 100 feet away, uphill from them. Their faces and torsos were covered with snow; only their hands and feet were sticking out. The wind was just screaming across the Col." The first body he came to turned out to be Namba, but Hutchison couldn't discern who it was until he knelt in the gale and chipped a three-inch-thick carapace of ice from her face. Stunned, he discovered that she was still breathing. Both her gloves were gone, and her bare hands appeared to be frozen solid. Her eyes were dilated. The skin on her face was the color of white porcelain. "It was terrible," Hutchison recalls. "I was overwhelmed. She was very near death. I didn't know what to do."

—JON KRAKAUER, *Into Thin Air:*
A Personal Account of the Mount Everest Disaster

Hillary Step: The most famous physical feature on Everest, the Hillary Step, at 28,750 feet, is a forty-foot spur of snow and ice. First climbed in 1953 by Edmund Hillary and Tenzing Norgay, the Hillary Step is the last obstacle barring access to the gently angled summit slope to the top of the world.

Cold Facts
Mount Everest is also called *Chomolangma*, meaning "Goddess Mother of Snows" in Tibetan and *Sagarmatha*, meaning "Mother of the Universe" in Nepalese. British surveyors named the peak for George Everest, a Surveyor General of India in the mid-nineteenth century.

Everest's current elevation is based on a GPS device implanted on the highest rock point under ice and snow in 1999 by an American expedition. The mountain is higher than 21 Empire State Buildings stacked on top of each other.

Mount Everest was dissected by glaciers into a huge pyramid with three faces and three major ridges on the north, south, and west sides of the mountain. Five major glaciers continue to chisel Mount Everest—Kangshung Glacier on the east; East Rongbuk Glacier on the northeast; Rongbuk Glacier on the north; and Khumbu Glacier on the west *and* southwest.

The summit temperature never rises above freezing. Summit temperatures in January average a minus 33°F and generally drop to a minus 76°F, or colder. These figures do not factor in wind chill.
> —"Mount Everest: The Highest Mountain in the World,"
> ThoughtCo.com, http://climbing.about.com/
> od/mountainclimbing/a/EverestFacts.htm

A depletion of oxygen, in addition to impairing thinking and physical action, diminishes the body's ability to manage cold.

Cold Facts
Antarctica is the earth's fifth largest continent, measuring 5.4 million square miles.

The thickest part of the ice sheet extends 16,000 feet into submerged basins in the rock.

Antarctica has the highest average elevation of any continent at 6,100 feet. The average elevation of North America is 2,300 feet.

The snowfall in Antarctica is so minimal that the continent has been called "the world's coldest desert." The interior receives less than one inch of precipitation a year, making it the driest continent on Earth.

Further: the Antarctic dry valleys in Victoria Land are among the driest places on Earth. Some scientists believe that no rain has fallen there for two million years.

Mean temperatures in the inland during the coldest month are from -40°F to -94°F and in the warmest month from 5°F to -31°F. At the coasts, the temperature ranges between 5°F to -22°F in winter and about 32°F in summer. On July 21, 1983, the Soviet station, Vostok, reported 128.6°F below freezing.

—see http://7summits.com/vinson/vinson.htm

Vinson Mastiff, at 16,050 feet, is the highest mountain in Antarctica.

> I have eaten
> the plums
> that were in
> the icebox
>
> and which
> you were probably
> saving
> for breakfast
>
> Forgive me
> they were delicious
> so sweet
> and so cold

—WILLIAM CARLOS WILLIAMS, "This Is Just to Say"

It was freezing cold, with a fog that caught your breath. Two large searchlights were crisscrossing over the compound from the watch-towers at the far corners. The lights on the perimeter and the lights inside the camp were on full force. There were so many of them that they blotted out the stars.

    With their felt boots crunching on the snow, prisoners were rushing past on their business—to the latrines, to the supply rooms, to the package room, or to the kitchen to get their groats cooked. Their shoulders were hunched and their coats buttoned up, and they all felt cold, not so much because of the freezing weather as because they knew they'd have to be out in it all day. . . . They went past the high wooden fence around the punishment block (the stone prison inside the camp), past the barbed-wire fence that guarded the bakery from the prisoners, past the corner of the HQ where a length of frost-covered rail was fastened to the post with heavy wire, and past another post where—in a sheltered spot to keep the readings from being too low—the thermometer hung, caked over with ice. Shukhov gave a hopeful sidelong glance at the milk-white tube. If it went down to forty-two below zero they weren't supposed to be marched out to work.

—ALEKSANDR SOLZHENITSYN,
One Day in the Life of Ivan Denisovich

He turned his attention to Beck, who lay twenty feet away. Beck's head was also caked with a thick armor of frost. Balls of ice the size of grapes were matted to his hair and eyelids. After clearing the frozen detritus from Beck's face, Hutchison discovered that the Texan was still alive, too: "Beck was mumbling something, I think, but I couldn't tell what he was trying to say. His right glove was missing and he had terrible frostbite. I tried to get him to sit up but he couldn't. He was as close to death as a person can be and still be breathing."

Horribly shaken, Hutchison went over to the Sherpas and asked Lhakpa's advice. Lhakpa, an Everest veteran respected by Sherpas and sahibs alike for his mountain savvy, urged Hutchison to leave Beck and Yasuko where they lay. Even if they survived long enough to be dragged back to Camp Four, they would certainly die before they could be carried down to Base Camp, and attempting a rescue would needlessly jeopardize the lives of the other climbers on the Col, most of whom were going to have enough trouble getting themselves down safely.

—JON KRAKAUER, *Into Thin Air:*
A Personal Account of the Mount Everest Disaster

At 4:35 P.M., Burleson was standing outside the tents when he noticed someone walking slowly toward camp with a peculiar, stiff-kneed gait.

"Hey, Pete," he called to Athans. "Check this out. Somebody's coming into the camp." The person's bare right hand, naked to the frigid wind and grotesquely frostbitten, was outstretched in a kind of odd, frozen salute. Whoever it was reminded Athans of a mummy in a low-budget horror film. As the mummy lurched into camp, Burleson realized that it was none other than Beck Weathers, somehow risen from the dead.
—Ibid.

Following his helicopter evacuation from the Western Cwm (the second-highest altitude helicopter rescue in human history!), Beck Weather's right arm was amputated halfway between the elbow and wrist. All four fingers and the thumb on his left hand were removed. His nose was amputated and reconstructed with tissue from his ear and forehead and he lost parts of both feet. He has since become a particularly effective motivational speaker . . . referring to himself as a "slow learner."

As a governor backing Alaskan oil interests, Sarah Palin sued the Department of the Interior for putting the polar bear on the endangered species list.

Permafrost underlies a quarter of the earth's surface. Permafrost is defined as any material that has remained frozen for two or more summers.

When a countrywide survey quizzed the public on what it means to be Canadian, the majority of respondents cited "not being American" as the primary characteristic of nationhood. As a secondary badge of identity, Canadians pointed to the existence of "our north." "We are a northern country," its citizens repeatedly asserted in the poll. "We have our Arctic." It is true that the Arctic occupies a higher percentage of the Canadian landmass than it does in any other country. Yet despite the central position of the Arctic in the national identity myth, only 100,000 out of 33 million Canadians reside north of 60° of latitude, the political boundary of the polar provinces (and roughly the tree line). Most are Inuit.
—SARA WHEELER, *Magnetic North: Notes from the Arctic Circle*

I settle myself in a crease in the tundra, out of the wind, arrange my clothing so nothing binds, and begin to study the far shore with the binoculars. After ten or fifteen minutes I have found two caribou. Stefansson was once asked by an Eskimo to whom he was showing a pair of binoculars for the first time whether he could "see into tomorrow" with them. Stefansson took the question literally and was amused. What the *inuk* probably meant was, Are those things powerful enough to see something that will not reach you for another day, like migrating caribou? Or part of a landscape suitable for a campsite, which you yourself will not reach for another day?

—BARRY LOPEZ, *Arctic Dreams*

By late August the film of grease ice that forms before the sea freezes was growing thicker every day: in the polar regions, the freeze and thaw of the sea replaces the rise and fall of sap.

—SARA WHEELER, *Magnetic North: Notes from the Arctic Circle*

The tracking of the Yukon River spring breakup at Dawson City, Yukon, Canada, dates back to the years of the Gold Rush. The average date of breakup based on 116 years of records is May 9 with a one month range between the earliest (April 29) and latest date (May 28). Recent dates for breakup have been May 3 in 2009, April 30 in 2010, and May 7 in 2011. Since 1896 people have held a lottery to guess the exact minute that the ice will go out in front of Dawson. A tripod holding a bell is placed on the river ice in front of the town. When shifting ice moves the tripod and rings the bell, breakup is official. Dawson City lies in a flood plain at the confluence of the Yukon and Klondike Rivers.

> The moon above the eastern wood
> Shone at its full; the hill-range stood
> Transfigured in the silver flood,
> Its blown snows flashing cold and keen,
> Dead white, save where some dark ravine
> Took shadow, or the somber green
> Of hemlocks turned to pitchy black
> Against the whiteness at their back.
> For such a world and such a night

Most fitting that unwarming light,
Which only seemed where'er it fell
To make the coldness visible.

—JOHN GREENLEAF WHITTIER, "Snow-Bound: A Winter Idyl"

After all, the *work norm* was senior in rank to the length of the work-day, and when the brigade didn't fill the norm, the only thing that was changed at the end of the shift was the convoy, and the work sloggers were left in the woods by the light of searchlights until midnight—so that they got back to camp just before morning in time to eat their dinner along with their breakfast and go out into the woods again.

There is no one to tell about it either. They all died.

And here's another way they raised the norms and proved it was possible to fulfill them: In cold lower than 60 degrees below zero, workdays were written off; in other words, on such days the records showed that the workers had not gone out to work; but they chased them out anyway, and whatever they squeezed out of them on those days was added to the other days, thereby raising the percentages. (And the servile Medical Section wrote off those who froze to death on such cold days on some other basis.) And those who were left who could no longer walk and were straining every sinew to crawl along on all fours on the way back to camp, the convoy simply shot, so that they wouldn't escape before they could come back to get them.

—ALEKSANDR SOLZHENITSYN, *The Gulag Archipelago*, III–IV

And on their feet the tried and true Russian "Lapti"—bast sandals—except that they had no decent "onuchi"—footcloths—to go with them. Or else they might have a piece of old automobile tire, tied right on the bare foot with a wire, an electric cord. (Grief has its own inventiveness. . . .) Or else there were "felt boots"—"burki"—put together from pieces of old, torn-up padded jackets, with soles made of a layer of thick felt and a layer of rubber. In the morning at the gatehouse, hearing complaints about the cold, the chief of the camp would reply with his Gulag sense of humor:

"My goose out there goes around barefoot all winter long and doesn't complain. . . ."

—Ibid.

But Stalin, in the execution of the broad brushstrokes of his hate, had weapons that Hitler did not have.

He had cold: the burning cold of the Arctic. "At Oimyakon [in the Kolyma] a temperature has been recorded of -97.8 F. In far lesser cold, steel splits, tyres explode and larch trees shower sparks at the touch of an axe. As the temperature drops, your breath freezes into crystals, and tinkles to the ground with a noise they call 'the whispering of the stars.'"

—MARTIN AMIS, *Koba the Dread: Laughter and the Twenty Million*

Kolyma is where the bulk of Russia's gold is. Over a span of two decades, starting in the early 1930s, more than two million prisoners were sent to mine that gold. The region quickly became legendary for its remoteness, its extreme cold, and its high death rate. Kolyma lies far beyond the reach of a railway line; getting there required crossing all of Russia on the Trans-Siberian and then traveling by ship for another week northward through the icy sea of Okhotsk. Hundreds of thousands of underfed Kolyma prisoners perished from malnutrition, from mining accidents, and from the bitter cold. . . . Kolyma is the coldest inhabited region on earth.

—JANUSZ BARDACH AND KATHLEEN GLEESON, *Man is Wolf to Man: Surviving the Gulag*

Blizzards could last not just for hours but days. The bodies of prisoners who got lost weren't found until springtime, often within one hundred meters of the zone. As a storm dispossessed the camp of its landmarks and reminders of lost freedom—the watchtowers, the work sites, the barbed wire—it also obliterated our sense of being locked in. Sometimes I fantasized that the blizzard would free us, that its power would wipe away the zone, the camp, all of Kolyma, and that when it was over we would be free.

—Ibid.

He walked very quickly over the snow crust, and when he reached the diner he found that he didn't want to go in yet. He would cross the highway and walk a little farther, then go into the diner to get warmed up on his way home. . . .

This is the very weather in which noses and fingers are frozen. But nothing felt cold.

He was getting close to a large woodlot. He was crossing a long slanting shelf of snow, with the trees ahead and to one side of him. Over there, to the side, something caught his eye. There was a new kind of glitter under the trees. A congestion of shapes, with black holes in them, and unmatched arms or petals reaching up to the lower branches of the trees. He headed toward these shapes, but whatever they were did not become clear. They did not look like anything he knew. They did not look like anything, except perhaps a bit like armed giants half-collapsed, frozen in combat, or like the jumbled towers of a crazy small-scale city—a space-age, small-scale city. He kept waiting for an explanation, and not getting one, until he got very close. He was so close he could almost have touched one of these monstrosities before he saw that they were just old cars. Old cars and trucks and even a school bus that had been pushed in under the trees and left. Some were completely overturned, and some were tipped over one another at odd angles. They were partly filled, partly covered, with snow. The black holes were their gutted insides. Twisted bits of chrome, fragments of headlights, were glittering.

He thought of himself telling Peg about this—how close he had to get before he saw that what amazed him and bewildered him so was nothing but old wrecks, and how he then felt disappointed, but also like laughing. They needed some new thing to talk about. Now he felt more like going home.

—ALICE MUNRO, "Fits"

Entr'acte

My great-grandfather was a sixteen year old living in gray-skied Copenhagen when he encountered the Mormon missionaries who converted him. Disowned by his family, he made his way, on his own, to America, and then on to Utah. His son, my grandfather on my father's side, left Utah for Montana during Prohibition. A Mormon bishop tipped him off that Federal Revenuers were on his trail for selling hooch to the Utes on the reservation. This same grandfather is reputed to have walked to Alaska to prospect for gold. When the permafrost made digging for ore overly difficult and costly, he humped his way back home to Butte, where instead of working in the Anaconda Company's copper mines, he played cards and boxed to make his living.

While attending The Montana School of Mines, I worked all my Christmas break at the Mt Con. On the 5200 level—one mile deep—the temperature was a high-humidity 90 degrees, or more. On the surface, at the mine's entrance, it was usually zero or below. All you wore were the clothes you were going to work in: one-piece union underwear and bib overalls. You shivered, shifting from one foot to the other, waiting your turn to enter the cage to be dropped underground. Three men would enter, facing out; four more stood front-to-back sideways; the final three would enter, facing in. Following payday, like clockwork, there was the sour odor of booze emanating through intermittent teeth into the space that was delivering us—and our stomachs—to the tropical clime one mile below the surface of what felt like a frozen planet.

> Glancing at the scene Sears thought of how the eighteenth- and
> nineteenth-century Dutch painters had cornered the skating scene
> and that before the values of the art market had become chaotic there
> were usually, at the end of the art auction, half a dozen unsold Dutch
> skating scenes leaning against the unsold umbrella stand beside the
> unwanted harpsichord. Brueghel had done some skating scenes but
> Sears had seen a skating scene—a drawing—from a much earlier
> period—the twelfth century he thought—and he always happily
> remembered Alan Gardener, the English paleontologist whose career
> was built on the thesis that the skate—or shate, since this came before
> any known language—had given Homo sapiens, as a hunter, the
> velocity that enable him to out strip Neanderthal man in the contest
> for supremacy. This was two hundred thousand years ago, much of
> the earth was covered with ice and the shate was made of the skull of
> the *Judas broadbill*. That Alan Gardener's thesis was all a fabrication
> was revealed very late in his career, but Sears found the poetry of his
> ideas abiding because the fleetness he felt on skates seemed to have the

depth of an ancient experience, and he had always been partial to any attempt to defraud the academic universe.

He put on his skates and moved off. This was quite as natural to him as swimming. He wondered why there were so few skaters on the ice and he asked a young woman. She was barely marriageable, with dark hair and gold rings in her ears, and she carried a hockey stick like a parasol. "I know, I know," she said, "but you see it hasn't been frozen over like this for over a century. It's been more than a century since it's been this cold without snow. Isn't it heavenly? I love it, I like it, I love it." He had heard exactly that exclamation from a lover so many years ago that he could not remember her name or the color of her hair or precisely what the erotic acrobatics were that they were performing.

—JOHN CHEEVER, *Oh What a Paradise It Seems*

> Sundays too my father got up early
> and put his clothes on in the blueblack cold,
> then with cracked hands that ached
> from labor in the weekday weather made
> banked fires blaze. No one ever thanked him.
>
> —ROBERT HAYDEN, from "Those Winter Sundays"

During the fall of 1993, a Midwestern university-trained anesthesiologist undertook a hike in the Rockies. In an unexpected turn, a boulder he rested against toppled to break his leg and to pin him to the earth. Unable to shift the stone, he said he thought of coyotes and wolverines and traps and bears. The doctor said later that he'd thought also of wild hogs and snakes, though he said he knew at the time that hogs were not native to the region and that snakes were in hibernation. "I wasn't myself," he explained.

Snow in the forecast, temperature and darkness falling, he determined to amputate his leg. Constructing a tourniquet of his shirt, the doctor then unfolded a fishing knife he'd sharpened the day before on a two-stage electric sharpener he'd ordered from Chef's Choice. He flayed the skin and fat to expose the left and then the right side of his knee joint. He parted tendons and ligaments, blood vessels, and nerves. He would work, he said, then pause to gather himself.

He severed the patellar ligament to separate the femur and tibia, then pulled the thighbone free. He dragged himself to reach his F250 Super Duty Ford 4WD.

It was a standard transmission he manipulated to the nearest town. U.S. Weather Service satellite photos confirmed a less than predicted storm—a light but wet snow—blanketed the area the following day. In this way, the doctor forgives himself for having removed his leg.

—DONALD ANDERSON, "Stumps"

Cold Facts

The oldest pair of skates known date back to about 3000 BC, found at the bottom of a lake in Switzerland. The skates were made from the leg bones of large animals, holes were bored at each end of the bone and leather straps were used to tie the skates on. An old Dutch word for skate is "schenkel," which means "leg bone."

Around the fourteenth century, the Dutch started using wooden platform skates with flat iron bottom runners. The skates were attached to the skater's shoes with leather straps. Poles were used to propel the skater. Around 1500, the Dutch added a narrow metal double-edged blade, making the poles a thing of the past, as the skater could now push and glide with his feet (called the "Dutch Roll").

In 1848, E. V. Bushnell of Philadelphia invented the first all-steel clamp for skates.

In 1865, Jackson Haines, a famous American skater, developed the two-plate all-metal blade. The blade was attached directly to Haines's boots. The skater became famous for his new dance moves, jumps, and spins. Haines added the first toe pick to skates in the 1870s, making toe pick jumps possible.

The first artificial ice rink (mechanically refrigerated) was built in 1876, at Chelsea, London, England and was dubbed the Glaciarium.

In 1914, John E. Strauss, a blade maker from St. Paul, Minnesota, invented the first closed toe blade made from one piece of steel, making skates lighter and stronger.

The largest outdoor ice rink is the Fujikyu Highland Promenade Rink in Japan, built in 1967 and boasts an ice area of 165,750 square feet— equal to 3.8 acres.

While covering the 1980 Winter Olympics in Lake Placid, *Washington Post* reporter Leonard Shapiro described American speed skater Eric Heiden as "a Secretariat on skates." But not even the legendary thoroughbred champion, accustomed to two-minute sprints, could likely follow the ice tracks laid down by Heiden over those nine days in February. The twenty-one-year-old Wisconsin native won five individual gold medals in speed skating. He won the speed-skating equivalent of a sprint, a marathon, and everything in between. He set five Olympic records and one world record, and defeated every single one of the 144 athletes competing in the five events. His medal haul began with a 0.34-second victory in the 500-meter race over Yevgeny Kulikov, the defending Olympic champion and world record holder. The next day he defeated world record holder Kai Arne Stenshjemmet in the 5,000-meter race by more than a second. Three days later he won the 1,000-meter race by 1.5 seconds, and forty-eight hours after that he captured his fourth gold, in the 1,500-meter race. The night before his final event, Heiden watched the US men's hockey team's "Miracle on Ice" upset of the Soviet Union in the tournament semifinals. The hockey team, which would later defeat Finland in the finals, won the only non-Heiden gold for the United States at the 1980 games. On the next-to-last day of competition, Heiden skated in the 10,000-meter race. He calmly and methodically smashed the world record by 6.2 seconds, winning in 14:28.13. His five gold medals remain a US Olympic record for a single winter games.

I have lived on this bridge for as long as I can remember, as long as I have been alive. Our house is newly built, for there was a fire on the bridge in 1212, just three years after the final work had been finished. The fire started on one side, and people rushed over from the other side to help quell the flames. But the wind was strong that day, and blew sparks across the bridge, catching the other end on fire, so that both ends were burning towards the middle, and the people on the bridge were trapped between the advancing walls of flames.

Three thousand bodies were plucked from the water below, the dead choosing to jump rather than burn, and dying all the same.

I think of that fire when I lie in bed at night. I can almost hear the hiss and whisper of the flame as it slinks across the bridge towards me.

But it is not fire that comes. It is ice. The winter is the coldest I can remember. The river shuts tight below me. I can see the men and oxen walking across it as though it were a road. Sometimes I can even hear the groans of the river under the ice, a groan like a dying animal, or a sleeping man.

I am lucky to live on the fourth arch, because when the ice started to break up, it shouldered through the archways and crashed against the starlings, and five of the arches towards the middle of the bridge collapsed. Ice is stronger than water, that is what I have learned. Ice is made from water, but it does not seem to remember water.

Five arches collapsed and houses were pitched from the bridge. Walls crumbled and timbers snapped. The bridge survived the fire, but it does not look as though it will be able to survive the ice.

London Bridge is falling down. Falling down. Falling down. London Bridge is falling down. My Fair Lady.

—HELEN HUMPHREYS, *The Frozen Thames*

In the 1600s, in Holland and Scandinavia (Sweden, Denmark, Norway, and Finland), people played a hockey-like game on ice. European settlers brought hockey with them to North America, where they found the Native Americans playing a game with sticks on a frozen pond. In Canada, hockey games were regular entertainment as early as the 1840s. After mass-produced ice skates went on sale in the 1860s, hockey became even more popular, with teams of twenty to thirty players trying to bat a puck through a goal that was made of two sticks jammed through holes in the ice. Players made their own sticks from tree branches until hockey sticks began to be mass-produced at the turn of the century. Original pucks were almost anything at hand, a pinecone, a rubber ball, a tin can. The first organized hockey games started in Kingston, Ontario, but it did not take long for the sport to organize throughout Canada. In 1881 two college students at McGill University created a rule book for play, and the game came to resemble modern hockey. In 1885 Kingston, Ontario, became the site of the first actual hockey league, the Amateur Hockey Association of Canada, with four teams. In the 1890s Americans became interested in Canada's favorite sport and formed a hockey league in New York City. Teams in Canada and the United States became professional organizations, that is, with paid players, in the late 1890s.

—"When Was Hockey Invented?" enotes.com, http://www.enotes. com/history/q-and-a/when-was-hockey-invented-286613

On November 1, 1959, Montreal Canadian all-star goalie Jacques Plante took a puck to the face where it sliced a gash from the corner of his mouth up through his nostril. After being stitched up, Plante refused to return to the ice unless his coach allowed him to wear a fiberglass mask he had designed himself for practices. Plante broke the way for other goalies in the NHL at a time when protective masks were considered a sissified approach to a "man's" game.

—"The night Plante made goaltending history," nhl.com, http://www.nhl.com/ice/news.htm?id=383063

At the highest levels of the game, a hockey puck can reach speeds of over 100 mph. The hardest slapshot ever recorded on radar was by Boston Bruins' defenseman Zdeno Chara at the 2011 NHL All-Star Skills Competition, where he recorded a velocity of 105.9 mph.

How would you like a job where when you made a mistake, a big red light goes on and 18,000 people boo?

—JACQUES PLANTE

The world speed record by a male human on skis is 156 mph. The women's downhill speed-skiing record is 124.9 mph. Speed skiing is a highly specialized aspect of downhill ski racing. It is an event where the purpose is to go in a straight line down the mountain as fast as possible. More common is downhill ski racing with turns. Technology changes and new materials, designs and innovations have all increased downhill speeds in competition to about 75 to

90 mph for men, and 60 to 75 mph for women. Depending on terrain, the average recreational skier swooshes along at an average of 15 to 20 mph.

The current world record for the longest ski jump is 246.5 meters: almost three football fields.

> The little town of Hagaru lay in a cleft in the mountains that otherwise appeared to the Marines to occupy the entire surface of North Korea. For those brief winter weeks of 1950, when it was occupied by the Americans, Hagaru resembled a nineteenth-century Arctic mining camp. Snow coated the peasant houses, the Marines' tents, the tanks and the trucks, the supply dumps and artillery pieces and command vehicles. The local sawmill was kept in perpetual motion by the engineers, cutting timber to strengthen positions and assist in the vital labor of airfield construction. Thin plumes of smoke from a hundred fires and stoves curled into the air on the rare days when the air was still. More often, they were whipped aside by the driving wind that stung every inch of exposed human flesh. At first, men marveled at the depths to which the thermometer could sink: -10, -14, -20 at night. Then they became as numb to the misery of the cold as to everything else. Many said that it was not only their capacity for physical activity that diminished, but even their speed of thought. General Smith himself found it increasingly difficult even to move his jaw to speak. The simplest action—loading a weapon, unbolting a steel section, rigging an aerial—became a laborious, agonizing marathon. The jeeps were kept running continually. In some cases their headlights were run on cables into key positions such as the sick-bay and operations tents to supplement the feeble Coleman lanterns. To start an engine required hours of work—thawing its moving parts, persuading its frozen oil to liquefy. Blood plasma froze. Medical orderlies were obliged to carry morphine Syrettes in their mouths to maintain their fluidity. For the men, the miraculously effective space heaters in the tents became the very focus of life. All this, before the enemy had even begun to take a hand.
>
> —MAX HASTINGS, *The Korean War*

The cold was if anything a more determined enemy than the Chinese.

It was pervasive and never let up, and if the natural cold registering on the thermometer up there on the Manchurian heights wasn't bad enough, most of the time they were in a kind of Manchurian wind tunnel where the cold had a constant extra bite to it. The men came to look like Ancient Mariners who had sailed too close to the North Pole, all of them bearded; their beards, filled with ice shavings, told the story. The cold made men want to quit and give up—made it hard to want to fight and live for another day—and yet every day they kept fighting. Years later, when one of the senior NCOs visited Chesty Puller at his home outside of Washington, Puller greeted him and said, "Hey, Sarge, thawed out yet?"

—DAVID HALBERSTAM, *The Coldest Winter:*
America and the Korean War

There are strange things done in the midnight sun
  By the men who moil for gold;
The Arctic trails have their secret tales
  That would make your blood run cold;
The Northern Lights have seen queer sights,
  But the queerest they ever did see
Was that night on the marge of Lake Lebarge
I cremated Sam McGee.

Now Sam McGee was from Tennessee, where the cotton blooms
    and blows.
Why he left his home in the South to roam 'round the Pole, God
    only knows.
He was always cold, but the land of gold seemed to hold him like a
    spell;
Though he'd often say in his homely way that "he'd sooner live in
    hell."

On a Christmas Day we were mushing our way over the Dawson
    trail.
Talk of your cold! through the parka's fold it stabbed like a driven
    nail.

If our eyes we'd close, then the lashes froze till sometimes we
    couldn't see;
It wasn't much fun, but the only one to whimper was Sam McGee.

And that very night, as we lay packed tight in our robes beneath the
    snow,
And the dogs were fed, and the stars o'erhead were dancing heel
    and toe,
He turned to me, and "Cap," says he, "I'll cash in this trip, I guess;
And if I do, I'm asking that you won't refuse my last request."

Well, he seemed so low that I couldn't say no; then he says with a
    sort of moan:
"It's the cursèd cold, and it's got right hold till I'm chilled clean
    through to the bone.
Yet 'tain't being dead—it's my awful dread of the icy grave that
    pains;
So I want you to swear that, foul or fair, you'll cremate my last
    remains."

A pal's last need is a thing to heed, so I swore I would not fail;
And we started on at the streak of dawn; but God! he looked
    ghastly pale.
He crouched on the sleigh, and he raved all day of his home in
    Tennessee;
And before nightfall a corpse was all that was left of Sam McGee.

There wasn't a breath in that land of death, and I hurried, horror-
    driven,
With a corpse half hid that I couldn't get rid, because of a promise
    given;
It was lashed to the sleigh, and it seemed to say: "You may tax your
    brawn and brains,
But you promised true, and it's up to you to cremate those last
    remains."

Now a promise made is a debt unpaid, and the trail has its own
    stern code.
In the days to come, though my lips were dumb, in my heart how I
    cursed that load.
In the long, long night, by the lone firelight, while the huskies,
    round in a ring,
Howled out their woes to the homeless snows—O God! how I
    loathed the thing.

And every day that quiet clay seemed to heavy and heavier grow;
And on I went, though the dogs were spent and the grub was
    getting low;
The trail was bad, and I felt half mad, but I swore I would not give
    in;
And I'd often sing to the hateful thing, and it hearkened with a grin.

Till I came to the marge of Lake Lebarge, and a derelict there lay;
It was jammed in the ice, but I saw in a trice it was called the "Alice
    May."
And I looked at it, and I thought a bit, and I looked at my frozen
    chum;
Then "Here," said I, with a sudden cry, "is my cre-ma-tor-eum."

Some planks I tore from the cabin floor, and I lit the boiler fire;
Some coal I found that was lying around, and I heaped the fuel
    higher;
The flames just soared, and the furnace roared—such a blaze you
    seldom see;
And I burrowed a hole in the glowing coal, and I stuffed in Sam
    McGee.

Then I made a hike, for I didn't like to hear him sizzle so;
And the heavens scowled, and the huskies howled, and the wind
    began to blow.
It was icy cold, but the hot sweat rolled down my cheeks, and I
    don't know why;

And the greasy smoke in an inky cloak went streaking down the
sky.

I do not know how long in the snow I wrestled with grisly fear;
But the stars came out and they danced about ere again I ventured
near;
I was sick with dread, but I bravely said: "I'll just take a peep inside.
I guess he's cooked, and it's time I looked"; . . . then the door I
opened wide.

And there sat Sam, looking cool and calm, in the heart of the
furnace roar;
And he wore a smile you could see a mile, and he said: "Please close
that door.
It's fine in here, but I greatly fear you'll let in the cold and storm—
Since I left Plumtree, down in Tennessee, it's the first time I've been
warm."

There are strange things done in the midnight sun
  By the men who moil for gold;
The Arctic trails have their secret tales
  That would make your blood run cold;
The Northern Lights have seen queer sights,
  But the queerest they ever did see
Was that night on the marge of Lake Lebarge
  I cremated Sam McGee.
                    —ROBERT W. SERVICE, "The Cremation of Sam McGee"

On the forward positions the cold was so appalling that an extraordinary improvisation, perhaps unique to this campaign, became necessary: the introduction of "warming tents," a few hundred yards behind
the front, where every two or three hours men retreated to thaw themselves a few degrees, to restore circulation to their deadened limbs, in
order that they might be capable of resistance when the Chinese came
again.

                    —MAX HASTINGS, *The Korean War*

And now there came both mist and snow,
And it grew wondrous cold:
And ice, mast-high, came floating by,
As green as emerald.

And through the drifts the snowy clifts
Did send a dismal sheen:
Nor shapes of men nor beasts we ken—
The ice was all between.

The ice was here, the ice was there,
The ice was all around:
It cracked and growled, and roared and howled,
Like noises in a swound!

—SAMUEL TAYLOR COLERIDGE, from
"The Rime of the Ancient Mariner"

As night fell on November 27, tens of thousands of Chinese soldiers came out of hiding, attacking American soldiers and Marines at all points around the Chosin Reservoir. The two companies dug into the west of Yudam-ni were freezing in make-shift foxholes when the overwhelming force attacked. In the darkness the Chinese swarmed the hill, coming within yards of the embattled Marines to toss grenades among them with deadly effectiveness. In one sector of the American perimeter, protected by two machine-guns, the horde quickly overran one of the key defensive positions. When a grenade landed near the only remaining machine-gun, Staff Sergeant Robert Kennemore stomped his foot on the grenade to push it into the snow, the subsequent blast throwing his body into the air. The Marines somehow held through the night. Sgt Kennemore was found, the stumps of his legs frozen in blood-caked snow, still alive, the severe cold having prevented a bleed-out.

—see "The Frozen Chosen," homeofheroes.com,
http://www.homeofheroes.com/brotherhood/chosin.html

"Frozen Chosin" was the site of the worst winter weather in Korea in 100 years. Add to that the surprise attack of some 120,000 Chinese soldiers . . .

The winter at Valley Forge generally conjures images of half-naked, starving soldiers battling the elements. This was not the case. This imagery is largely the result of early, romanticized interpretations of the encampment story which were meant to serve as a parable about American perseverance.

— "American Revolution: Winter at Valley Forge,"
militaryhistory.about.com, http://militaryhistory.about.com/
od/battleswars16011800/p/valleyforge.htm

The cold was worse on the clear nights. They camped on a flat place beside a river, almost within sight of the enemy on the other side. As wretched as the days were because of the cold and the fear and the sickness, the night was terror. The rivers froze, and at night the Chinese or the North Koreans would inch their way across it, one or two at a time, and do their killing with knives. It was legend, those killings, designed to terrify.

That night, or maybe it was morning, an assassin crept into his group, as he slept, and killed a man just inches away. My daddy reached out to shake him, and felt the blood that had leaked from his neck.

He scrambled out of the shelter and into the biting cold, and saw him, the killer, on the ice.

The man lay flat on his belly, to keep the ice from breaking, and slithered and squirmed like some kind of slow-moving reptile, just a few feet from the bank. My daddy ran down to the river's edge and, unthinkingly, straight out onto the ice, slipping down hard on his hands and knees, hearing the ice crack. But he lunged forward and grabbed the man. They fought, frantic, crazy. My daddy must have lost his rifle because he never mentioned using it, and if he pulled his knife he never said. He knew the other man had a knife, had to have one, but my daddy did not see it. Maybe, in his rage, his terror, he did see it and didn't care.

Finally he fought to his knees and pushed himself on top of the man, and the ice popped and cracked again. The other man clawed at my daddy's face, screaming, and finally fought free. He tried to do the impossible, to walk on that thin ice, and plunged straight through it.

The man rose up, his hands clutching the edge, and although he didn't know it, he was already dead. The cold, that unbearable cold, would take him even if he could get out of the water. My daddy could have left him that way, could have let the ice have him. Instead, he reached down into the water and put both hands on top of the man's head and pushed him down, again and again, till there was no need anymore.

When the thing was done, my daddy, freezing, crawled back to where he guessed dry ground was, to the shelter, peeled off his wet, freezing gloves and shoved both hands between his legs to warm the numbness away.

If there was satisfaction in what he had done, he didn't say.

—RICK BRAGG, *All Over but the Shoutin'*

Entr'acte

Fifty years and the picture sticks: my father knocking icicles from the eaves with a broom. It's the night before Christmas in Butte, Montana, where, in 1959, the miners are on strike, and here's my long-legged father, out of work, knocking on ice. My father has jerked the broom from a snowbank beside the carport—the carport he'd built the summer before, with salvaged 4×4 posts and corrugated Fiberglas. The Fiberglas is a mint green that tints sun.

My father is sweeping icicles because we are about to make ice cream. It's damned cold, which is not the way I would have put it then, little Mormon boy that I was. When I was a kid in Butte, winter temperatures often fell to 30, 40, even 50 below. It's cold and my father and I are in our sheepskins and lined buckle-up overshoes. Snow is at our knees. When the icicles snap, they drop with intent. My father has told me to *Pay Attention Now. Stay Tuned.*

I checked the street. What if somebody saw? What would they think of someone constructing ice cream in the dead of winter here, in the dark? And if we were going to do that, why couldn't we drive to a gas station to buy block ice? We have ice, my father said. Perhaps it was best to use the ice at hand, for we'd have driven to the gas station in the 1949 two-tone—black-and-white—Hudson Hornet my father had formed with an axe and a hacksaw into what he thought passed as a half-ton truck. After he'd removed the back seat to connect the space with the trunk, he folded the roof to contain the front seat. He'd bent the roof by ramming it like a fullback, a pulling guard. I was upright in what had been the back seat. You helping? he asked. We rammed the roof.

He hacked a hole for a Plexiglas rear window and metal-screwed everything as tight as he could. During winter, the heater had to hump to keep the cab heated. The finishing touch was quarter-inch sheet metal welded to the frame for the truck bed. There was no tailgate.

—DONALD ANDERSON, "Rock Salt"

Under the direction of Washington's army engineers, his wintering soldiers constructed over 2,000 log huts laid out along military streets. In addition, defensive trenches and five redoubts were built to protect the encampment.

—"American Revolution: Winter at Valley Forge,"
militaryhistory.about.com, http://militaryhistory.about.com/
od/battleswars16011800/p/valleyforge.htm

Winter fell onto us like a cold, suffocating weight. I woke up one morning to see high, feathery cirrus clouds blowing fast across the sky. By noon the next day, the empty, blue skies of autumn had been replaced by a leaden gray. That night it started snowing, covering the valley with a deadening blanket of white. Then the cold came and it was like some kind of monster, crawling into your lungs, settling in your bones. We huddled around our kerosene heater, and the stink of it permeated everything. There was kerosene in the food, in the water. The world filled up with that sour chemical taste.

—JOHN HAGGERTY, "Tumbleweeds"

That winter was a bad one. It would have been hard without the war, but the combination of cold and combat was disastrous. At first, we would see little bands of travelers, but as the grim weather kept on, we saw fewer and fewer people. Temperatures had been below freezing for two straight weeks when we found the first tumbleweed. There had been a refugee party passing to the east a day earlier, and I guess he had been with them. I called Atkins over when I found him. An old guy in shabby cotton clothes, curled up, frozen and stiff. I found myself getting angry at him—this pathetic, frozen, gray little man. What right did he have to come and die up against our fence? Why did he pick us, when there were so many other places in this terrible country he could have done it? Death was everywhere. Death was the air we breathed and the water we drank. Why did we have to have one more reminder?

—Ibid.

In January, I skated out onto Lake Erie, which that year was frozen nearly to Canada. I stared at its ominous expanse. I left the shore one evening on my hockey skates, a wool cap pulled over my ears and a long scarf wound around my neck and crisscrossed over my chest beneath my blue navy-surplus pea jacket. I meant to learn courage out on the ice, to avoid the specter of cowardice by skating all the way either to Canada or, if the icebreaker had been through, to the Livingstone ship channel. I struggled over the corrugations of the near-shore ice, then ventured onto glassier black ice that rewarded me with long glides between strokes of my hollow-ground blades. Bubbles could be seen and, occasionally, upended white bellies of perch and rock bass, as the sheen of glare ice, wide as my limited horizon, spread east toward Ontario; I dreamed of landing on this foreign shore, from whence the redcoats once launched sorties against our colonial heroes. I would tell Mrs. Andrews what I had done. Reading schoolbooks had embittered me against the British and the American South, while my uncles handled the job for Germany and Japan. I meant to visit the old British fort Amherstburg and skate home with tales of imperial ghosts and whatever other secret existences I might discover in places where no human is expected.

—THOMAS MCGUANE, "Ice"

Now the sound of my blades, which had seemed to fill the air around me, was replaced by another as murmurous as a church congregation heard from afar. I glided toward the sound when suddenly a vast aggravation of noise and turbulence erupted as a storm of ducks took flight in front of me; it was water. I heard the ominous heave of the lake. I turned to skate straight away—or not quite straight, because after some minutes of agitated effort I found myself at water's edge again, water sufficiently fraught that it had broken back the edge of ice, heaving it in layers upon itself. I skated away from that too, and when once more surrounded by darkness and standing squarely on black ice, I stopped and recognized that I was lost. I was suspended in darkness.

—Ibid.

Oliver wasn't bad at all on the next visit to Rikers. The weather was colder and he got to wear his favorite Spider-Man sweater, which Boyd said was very sharp.

"Your Mom's looking good too," Boyd said to Oliver.

"Better than Lynette?"

I hadn't meant to say any such whiny-bitch thing; it leaped out of me. I was horrified. I wasn't as good as I thought I was, was I?

"Not in your league," Boyd said. "Girl's nowhere near." He said this slowly and soberly. He shook his onion head for emphasis.

The rest of the visit went very well. Boyd suggested that Oliver now had the superpower to spin webs from the ceiling—"You going to float above us all, land right on all the bad guys"—and Oliver was so tickled he had to be stopped from shrieking with glee at top volume.

"Know what I miss?" Boyd said. "Well, that, of course. Don't look at me that way. But I also miss when we used to go ice-skating."

We had gone exactly twice, renting skates in Central Park, falling on our asses. I almost crushed Oliver one time when I went down. "You telling everyone you're the next big hockey star?" I said.

"I hope there's still ice when I get out," he said.

—JOAN SILBER, "About My Aunt"

In stories of the Great Lakes
November gales freeze sailors

where they stand, fingers
knotted in ice around the rigging,
wheel house sheathed in frozen rain.
Weighted with this cargo of ice
the ship groans, and sheets of isinglass
crack in the shudder of being hit
with a cavernous wave. Ore shifts
in the hold; the lake freezes on deck,
layer after layer laid down
until steel bends, and rivets pop like a bone
ripped from the mooring of its socket.

—DALE RITTERBUSCH,
from "The Monuments We Swear By"

He sprung from the cabin window . . . upon the ice-raft which lay
close to the vessel. He was soon borne away by the waves, and lost in
darkness and distance.

—MARY SHELLEY, *Frankenstein*

Attila Petschauer was a Hungarian winner of three Olympic fencing
medals, including two gold. Swept up in the German occupation of
Hungary during World War II, he was protected for a while from his
Jewishness by his reputation as a celebrated sportsman and accorded
a special "document of exemption." During a routine check of identi-
fication, however, Petschauer found he had left some of his "papers"
at home—an unacceptable explanation to his Nazi inquisitor. He was
deported to a labor camp in the Ukraine. During a lineup of prisoners,
Petschauer was recognized by Lieutenant Colonel Kalman Cseh, who
had been an equestrian competitor for Hungary in the 1928 Olympics.
Though the two had once been friends, Cseh exhorted camp guards
to taunt his one-time comrade. "The guards shouted: 'You, Olympic
fencing medal winner . . . let's see how you can climb trees.'" It was
midwinter and bitter cold. Petschauer was ordered to undress, then
climb a tree. The amused guards then ordered him to crow like a
rooster while they sprayed him with water. Frozen from the water, the

unclothed Petschauer died. His parents had named him for that other
Hungarian half-dressed sportsman, Attila the Hun.

—see "Attile Petschauer," International Jewish Sports Hall of Fame,
http://www.jewishsports.net/BioPages/AttilaPetschauer.htm

Yes, there are certainly icicles hanging by the wall, down the clap-
boards, dripping from the eaves, downspouts, gutters and a last course
of curling shingles. Thirty days. A month—but who's counting—of
days below the freeze mark; it never stops. As for a merry note,
whacking stalactites with a shovel does nothing but break the peace:
those sharp, cutting shards explode upon the walk though better
there than on the heads of those braving it out to pick up the morning
paper, ice sheets on the steps, swords of ice overhead—a dangerous
place despite the serenity of winter. Almost nothing moves in this
frigid pallor of air. Even hatred freezes: would it were so in the headlines—another air strike, more collateral damaged. I swing hard at a
thick, glacial flow of ice twisting the gutter overhead. It shatters, and
the headlines break: just another day of cold and damming ice. And
fire, always fire, somewhere, on the other side of this world.

—DALE RITTERBUSCH,
"After Shakespeare's 'When Icicles Hang by the Wall'"

In 1845, Sir John Franklin and 138 officers and men embarked from
England to find the northwest passage across the high Canadian
Arctic to the Pacific Ocean. They sailed in two three-masted barques.
Each sailing vessel carried an auxiliary steam engine and a twelve-
day supply of coal for the entire projected two or three years' voyage.
Instead of additional coal, according to L. P. Kirwan, each ship made
room for a 1,200 volume library, "a hand-organ, playing fifty tunes,"
china place settings for officers and men, cut-glass wine goblets, and
sterling silver flatware. The silver was of ornate Victorian design, very
heavy at the handles and richly patterned. Engraved on the handles
were the individual officers' initials and family crests. The expedition
carried no special clothing for the Arctic, only the uniforms of Her
Majesty's Navy.

The ships set out in high dudgeon, amid enormous glory and fanfare. Franklin uttered his utterance: "The highest object of my desire is faithfully to perform my duty." Two months later a British whaling captain met the two barques in Lancaster Sound; he reported back to England on the high spirits of officers and men. He was the last European to see any of them alive.

—ANNIE DILLARD, "An Expedition to the Pole"

The winter of 1996–97, combined with the 1997 spring floods, caused the worst natural disaster in recent history for North Dakota, eastern South Dakota, and western Minnesota. Above-normal snowfall in central and eastern North Dakota during the winter and an April blizzard, caused the worst flooding in the Red River of the North and Missouri River Basins in more than 100 years. The heaviest snowfalls occurred along the main stems of the Red River of the North and the Missouri River and were about 300 percent greater than normal. About 117 inches of snow were recorded in Fargo, 96 inches in Grand Forks, and 101 inches in Bismarck. Elsewhere in the region, snowfalls were well above seasonal averages. Melting of the snowpack and thawing of ice began in late March on rivers and streams in the southern and western parts of the State. Flows were inhibited by a blizzard that occurred on April 5–6, 1997. The blizzard brought a severe drop in temperatures, winds up to seventy miles per hour, and up to two feet of snow with drifts many feet higher in several areas. In southeastern North Dakota, the blizzard was preceded by wind-driven rain and sleet. The wind and ice toppled trees and power lines, leaving thousands of people without power for days. Thousands of people were forced to flee their homes, some permanently, as floodwaters and severe weather caused over $5 billion in damage to the region.

—see "Red River of the North Flooding—1997,"
North Dakota Water Science Center,
http://nd.water.usgs.gov/photos/1997RedFlood/

The Red River of the North is one of the few rivers in the United States to flow directly north into Canada. The basin flood plain lies in a glacial lakebed and is relatively flat (less than 0.5-foot drop in elevation per mile in the reach downstream from Grand Forks, North Dakota). Because of the flat basin, the shallow river channel, and the northerly flow, the timing of spring thaw and snowmelt can greatly aggravate flooding in the basin. Snow and ice in the headwaters of the Red River of the North begin to melt first, when areas downstream remain largely frozen. The melt pattern can cause ice jams to form, and substantial backwater can occur as flow moves northward toward a still-frozen river channel.

—Ibid.

On April 6, 1997, the stage of the Red River of the North at Wahpeton was 19.42 feet. On April 17, 1997, the peak stage of the Red River of the North at Fargo, about 96 river miles north of Wahpeton, was 39.57 feet. On April 18, 1997, the peak stage at Fargo was 39.72 feet.

High flows continued to move downstream in the Red River of the North. On April 18, 1997, the peak stage of the Red River of the North at Grand Forks was 52.04 feet, and the peak flow was 137,000 cubic feet per second. The peak flow was unusual because it resulted from the convergence of flows from the Red Lake River in Minnesota, flows from the main channel, and breakout flows from the Red River of the North that were conveyed by old Red River of the North oxbows. Breakout flows occurred upstream from Grand Forks when plugs in the upstream end of the oxbows either were overtopped or washed away, which caused a flow of about 25,000 cubic feet per second to arrive at the confluence of the Red Lake River and the Red River of the North at Grand Forks. The flow of 25,000 cubic feet per second coincided with the peak flow of the two rivers. On April 24, 1997, the peak stage at the Red River of the North at Drayton was 45.55 feet, and the peak flow was 124,000 cubic feet per second. On April 27, 1997, USGS personnel measured 141,000 cubic feet per second in the Red River of the North at Pembina, which is located about 2 miles upstream from the international boundary with Canada.

—Ibid.

Years later, civilization learned that many groups of Inuit—Eskimos— had hazarded across tableaux involving various still-living or dead members of the Franklin expedition. Some had glimpsed, for instance, men pushing and pulling a wooden boat across the ice. Some had found, at a place called Starvation Cove, this boat, or a similar one, and the remains of the thirty-five men who had been dragging it. At Terror Bay the Inuit found a tent on the ice, and in it, thirty bodies. At Simpson Strait some Inuit had seen a very odd sight: the pack ice pierced by the three protruding wooden masts of a barque.

For twenty years, search parties recovered skeletons from all over the frozen sea. Franklin himself—it was learned after twelve years—had died aboard ship. Franklin dead, the ships frozen into the pack winter after winter, their supplies exhausted, the remaining officers and men had decided to walk to help. They outfitted themselves from the ships' stores for the journey; their bodies were found with those supplies they had chosen to carry. Accompanying one clump of frozen bodies, for instance, which incidentally showed evidence of cannibalism, were place settings of sterling silver flatware engraved with officers' initials and family crests. A search party found, on the ice far from the ships, a letter clip, and a piece of that very backgammon board which Lady Jane Franklin had given her husband as a parting gift.

—ANNIE DILLARD, "An Expedition to the Pole"

Now is the winter of our discontent . . .

Because our name is a town in their country, the two Dutch exchange research chemists called for my sister each morning to skate across the lake. They were insistent and prompt; as a compatriot she had to go.

Wisconsin is cold at five in the morning, and that winter was very long. Lake Mendota froze in November and still supported cars in late March. They had to dodge those cars, driven mid-lake by fishermen who huddled over ice holes and were startled by their flashlights when they skated past before dawn.

She never made it across. They started flanked, then my sister dropped from the center and halfway was two blocks behind, lingered at the fishing holes, tried to warm up at the stoves, talk to the men in grunted exchanges that broke their attention on their lines. By December she headed straight for the cars and asked to wait inside, where she dozed with her blades braced on dashboards or, breathing a hole in the window frost, watched for the Dutchmen's return. She could just make them out: two black figures in flapping coats, bent like grackles, growing larger against the far shore. They'd collect her and see her home and say she'd do better next day.

One morning she stopped. It was after a night of study, she had an exam, swollen ankles, and didn't answer the bell. Their knocking roused the housemother, who pulled her by the robe to the door. She couldn't face them, only yelled through the mail slot, I can't, I forgot to tell you: I'm half Serb! They skated alone till thaw. Soon after, they went home.

Aging, she wakes early. It's just as cold. Stops only for coffee. Laces up, heads across the lake. She makes it in twenty minutes, less when the wind shifts. She'd like them to know.

—DIANE VREULS, "Let Us Know"

To separate *Now is the winter of our discontent* is to do a disservice to Richard III's intention as he continues in this opening scene of the play: "Made glorious summer by this son of York / And all the clouds that low'r'd upon our house / In the deep bosom of the ocean buried." Nonetheless the opening phrase is, more often than not, taken out of context. Even Steinbeck took it as the title of his final novel.

Some juvenile delinquents had heaved a brick through the windshield on the drivers' side, so the cold and snow tunneled right into the cab. The heater didn't work. They covered themselves with a couple of blankets Kenny had brought along and pulled down the muffs on their caps. Tub tried to keep his hands warm by rubbing them under the blanket but Frank made him stop.

They left Spokane and drove deep into the country, running along black lines of fences. The snow let up but still there was no edge to the land where it met the sky. Nothing moved in the chalky fields. The cold bleached their faces and made the stubble stand out on their cheeks and along their upper lips. They stopped twice for coffee before they got to the woods where Kenny wanted to hunt.

Tub was for trying someplace different; two years in a row they'd been up and down this land and hadn't seen a thing. Frank didn't care one way or another, he just wanted to get out of the goddamned truck. "Feel that," Frank said, slamming the door. He spread his feet and closed his eyes and leaned his head way back and breathed deeply. "Tune in on that energy."

"Another thing," Kenny said. "This is open land. Most of the land around here is posted."

"I'm cold," Tub said.

Frank breathed out. "Stop bitching, Tub. Get centered."

"I wasn't bitching."

"Centered," Kenny said. "Next thing you'll be wearing a nightgown, Frank. Selling flowers out at the airport."

"Kenny," Frank said, "you talk too much."

"Okay," Kenny said. "I won't say a word. Like I won't say anything about a certain babysitter."

"What babysitter?" Tub asked.

"That's between us," Frank said, looking at Kenny. "That's confidential. You keep your mouth shut."

Kenny laughed.

"You're asking for it," Frank said.

"Asking for what?"

"You'll see."

—TOBIAS WOLFF, "Hunters in the Snow"

> The temper of
> water waiting for
> its shape in
>
> the unrelenting
> rush of things
> in their freezing.
>
> —PAUL HOOVER, from "Winter (Mirror)"

> . . . There is
> nothing the sun
> cannot explain . . .
>
> —Ibid.

Intense cold in the upper atmosphere of the Arctic last winter activated ozone-depleting chemicals and produced the first significant ozone hole recorded over the high northern regions, scientists have reported in the journal *Nature*.

While the extent of the ozone depletion is considered temporary, atmospheric scientists described it as a striking example of how sudden anomalies can occur as a result of human activity that occurred years ago.

"The root cause is the residual products from the CFCs that were released throughout the 20th century," said Michelle L. Santee, a planetary scientist at NASA's Jet Propulsion Laboratory and one of the paper's authors. "But they are very long-lived, and it will take a few decades for them to be cleansed from the atmosphere."

—*THE DENVER POST*, October 4, 2011

"The goal is to trick the organ into thinking it never left the body," Maximilian Polyak, organ preservationist at New York Presbyterian Hospital, explains. Ice cuts down on cell swelling, which could rupture cell membranes, and "drastically" lowers the organ's rate of metabolism, reducing its need for oxygen. "For every 10° [C; 18°F] drop below normal organ temperature," Polyak says, "there's a four-fold drop in metabolic rate. By the time the temperature is down 4° [C; 39°F], the organ's metabolism is only one-fourteenth to one-sixteenth of what it is normally." The organ is quiescent and bloodless and deteriorating yet still "very much alive." A donor kidney can be kept on ice for 48 to 72 hours, a liver for 24 to 36 hours, a heart for four to eight hours.

—MARIANA GOSNELL,
Ice: The Nature, the History, and the Uses of an Astonishing Substance

Entr'acte

Warned not to attach my tongue to metal in Montana during winter, I began eyeing the galvanized posts on the grounds of a house I passed on my way to

Emerson Elementary. The mercury registered thirty below when I chaperoned my experiment.

> She walks along the crest of the hill and stops. Behind her there is a curve of footsteps arching along the hill and back down to the house. In front of her there is a large stretch of virgin snow. She takes a leaping giant step forward and then sits down and leans back carefully into the snow, keeping her feet together and her arms by her sides, so that she is lying quite straight and looking up at the soft shadow that curls along the face of the full moon. She begins to move, pushing her arms out across the snow and bringing them up toward her head, then down to her sides. She moves her legs across the snow in a scissors motion—out, then back.
> Marguerite, aged twenty-nine, mother of two, is making a snow angel.
> In a moment she will be so cold that she will have to stand up and go home. She will have to scuffle around on top of the angel to obliterate the evidence of this whimsy. But now she lies deep in the snow and moves her arms and legs very slowly. She moves with the slow rhythm of the moon moving across the sky. She moves with the slow beat of the stars pulsating their light to stars in other galaxies. She has a pair of white wings and a white skirt. She has white moonlight and the clean white frost of her own breath, and now, alone on the hillside in the white universe, with the shadow of her own footsteps reaching back to the house like a lifeline, Marguerite feels the calm of a great and voluptuous sigh.
>
> —STEPHANIE VAUGHN, "Snow Angel"

The ozone layer lies approximately 10–25 miles above the Earth's surface, in the stratosphere. Depletion of this layer by ozone depleting substances will lead to higher UVB levels, which in turn will cause increased skin cancers and cataracts and potential damage to some marine organisms, plants, and plastics.

> —see "Ozone Layer Protection," United States Environmental Protection Agency, http://www.epa.gov/ozone/defns.html#uvb

"It was Christmas Day in Denver, brother." And Manny told us of sitting at the bar just after noon sipping a V. O. and ginger, the only thing he ever drank. He was drinking alone, thinking about business, about people who owed him money and how hard it was to collect at Christmastime. He's stepped into the bar because the air in the streets was so cold it "froze my face, brother. You know, it hurt your skin."

Curtis nodded and began talking about winters he'd known. In the corner of the mess hall, the kleptomaniac was singing "Blue Eyes Crying in the Rain," his voice high and plaintive, and Dozer was laughing too hard as he won a hand, and Manny cut Curtis off about the ice storm that had sealed a canyon in Curtis's youth. Manny kept talking: "It's Christmas, but the bar is full, man, full of sad-assed players like me." It was warm and dark as a cave, and Freddy Fender was on the jukebox, and the bartender was an older woman with big breasts she didn't mind showing off in a low-cut sweater, two Santa earrings swinging at each ear. Manny ordered another V. O. and ginger and was fishing in his pocket for some cash when a man behind him jumped up from a table and ran outside. From where he sat, Manny could see the whole scene out the oval window of the front door, as if it had been framed like that just for him. And he knew the man who'd rushed out to the street. It was Little Junior, a punk in the neighborhood who was into everything but his own business. He was small and always packing heat, and now he was pointing his finger an inch from a big man's face, a black man Manny had seen around for years. Manny turned to the barmaid who'd just finished mixing his drink, setting it on the bar in front of him, when four shots thumped through the air outside and Manny turned to see through the oval frame Little Junior falling away, the black man out of sight.

"We all went out there. You know, we stood around Little Junior just looking at him 'cause he was gone, brother. You didn't have to take his pulse or nothing."

Little Junior lay flat on his back, his arms and legs spread like he was going to make a snow angel. Only there was no snow, just the frozen air, and all four shots had ripped through his chest and now

Manny was getting at the end of his story, the point that made him tell us in the first place.

"There was steam rising out of them holes, man. You could see it coming out of him." Manny looked from Curtis to me. He shook his head. "I know that was the heat of his body, but it was Christmas Day, brother, so I seen that as his soul, Little Junior's dirty little soul, rising up over us all."

—ANDRE DUBUS III, *Townie*

And here comes the snow, a language
in which no word is ever repeated.

—WILLIAM MATTHEWS, from "Spring Snow"

They say that every snowflake is different. If that were true, how could the world go on? How could we ever get up off our knees? How could we ever recover from the wonder of it?

—JEANETTE WINTERSON

No snowflake ever falls in the wrong place.

—ZEN PROVERB

My father thought that everything could be repaired with duct tape. He used so much of it to cover rust holes on our station wagon that on the days when he picked me up from school I would jump into the car and immediately slide down out of sight. I had a Seattle Seahawks coat from the Sears catalog with glossy silver arms that tore with wear, and my father couldn't have been more pleased that his duct tape matched. He used it to patch my silver moon boots too. I was horrified by the amount of duct tape I sported in winter. I looked like a ragged astronaut as I slowly forged my way through the snow to the ruins of the old Poolville dam. I had been warned to stay away from it, but I wandered there anyway. The town was a different place in winter, and I felt different in it. It was the time of year when you were conscious of the act of dressing for the weather. You had to wear protection. I understood that the elements were against me, but I was not old enough to be afraid of what that meant.

—BENJAMIN BUSCH, *Dust to Dust: A Memoir*

Clumps of slush passed as the river moved toward the oceans. It was during cold like this that snow fell into the river and became pasty, the motion of water creating a temperature that kept its volume from freezing but did not allow the snow traveling on it to melt. There were no individual flakes visible, just like ice and water betrayed nothing divisible from their sum, and there was no making of time in the passing of water. The river took on a deadly hue and an opacity that hid its rocky bed from view. I wondered how the color of water could change, disguise its depth like paint. I understood when it browned with mud during floods, but this seemed to be a color without particles or pigments. It could secretly be as deep as the center of the sea.

—Ibid.

. . . my attention was caught by a snowflake on my coat-sleeve. It was one of those perfect little pine trees in shape, arranged around a central spangle. This little object, which, with many of its fellows, rested unmelting on my coat, so perfect and beautiful, reminded me that Nature had not lost her pristine vigor yet, and why should man lose heart?

—HENRY DAVID THOREAU

No snowflake in an avalanche ever feels responsible.

—VOLTAIRE

Is it better to die in winter? The question
arose today in my high school history
lecture.—Millions of green leaves fell
when Germany invaded Russia
in the coldest moments of World War II.
Excitement built when I said, "Russians drove
tanks built in Siberian factories; bare of paint,
driven by men who made them, these tanks hit
Hitler's men hard." My students enjoyed hearing
how at Stalingrad, Stalin's snipers thinned
German officers' ranks. Excitement built again
in the males, as I told how across the front, Russia's
T-34 tanks destroyed German Panzer divisions,

overwhelmed German soldiers in the allied race
to Berlin. Ukrainians welcomed Germans,
then Russians. One male student laughed
when I said, "Captured German soldiers picked
corn from excrement to survive, fought
over tiny seeds until they had no strength, and fell
in the snow." It's always the same when I tell
the story. As I looked out the window on my left,
my European history class discussed problems
Germany had with two fronts in both wars. Outside
the classroom, a wind rattled the trees—I thought
about my brother all boxed up from Iraq
two years ago. He died in the desert.
Green leaves fall, keep falling.

—GREGORY STENTA, "Lecture on Stalingrad"

Annually, Antarctica produces some five thousands bergs, about
6.5 times the production of the Arctic. The average size of Antarctic
bergs is much greater than that of Arctic bergs, each Antarctic berg
averaging about one million tons of pure fresh water. Total production equals nearly 690 cubic kilometers of ice. Unlike Greenland
bergs, calving off fast-moving glaciers, Antarctic bergs tend to calve
from ice shelves or from the tongues of outlet glaciers protruding
into the sea. The Greenland bergs, accordingly, resemble small
peaks, while the Antarctic bergs resemble great tabular plateaus. . . .
One sighted in 1927 was reported as 160 kilometers long, with a free-
board height of 35 meters. Others have been measured at 140 × 60
kilometers, 100 × 70 kilometers, and 100 × 43 kilometers. The greatest, tracked in 1965, was 140 kilometers long and featured a surface
area of 7,000 square kilometers. One colossal berg, the Trolltunga,
began as a severed ice tongue roughly the size of Belgium.

—STEPHEN J. PYNE, *The Ice: A Journey to Antarctica*

But, after all, it is not what we see that inspires awe, but the knowledge of what lies beyond our view. We see only a few miles of ruffled
snow bounded by a vague wavy horizon but we know that beyond that

horizon are hundreds and even thousands of miles which can offer
no change to the weary eye, while on the vast expanse that one's mind
conceives one knows there is . . . nothing but this terrible limitless
expanse of snow. . . . Could anything be more terrible than this silent,
wind-swept immensity when one thinks such thoughts?

—ROBERT SCOTT, *The Voyage of the Discovery*

This is the period between life and death. This is the way the world
will look to the last man when he dies.

—RICHARD BYRD, *Alone*

Permafrost, the permanently frozen soil that underlies the tundra,
presents arctic trees with still other difficulties. Though they can penetrate this rocklike substance with their roots, deep roots, which let
trees stand tall in a windy landscape, and which can draw water from
deep aquifers, serve no purpose in the Arctic. It's too cold to stand
tall, and liquid water is to be found only in the first few inches of soil,
for only this upper layer of the ground melts in the summer. . . . Trees
in the Arctic have an aura of implacable endurance about them. A
cross-section of the bole of a Richardson willow no thicker than your
finger may reveal 200 annual growth rings beneath the magnifying
glass. Much of the tundra, of course, appears to be treeless when, in
many places, it is actually covered with trees—a thick matting of short,
ancient willows and birches. You realize suddenly that you are wandering around on top of a forest.

—BARRY LOPEZ, *Arctic Dreams*

I remember the cold night you brought in a pile of logs and a chipmunk jumped off as you lowered your arms. "What do you think
you're doing in here?" you said, as it ran through the living room.
It went through the library and stopped at the front door as though
it knew the house very well. This would be difficult for anyone to
believe, except perhaps as the subject of a poem. Our first week in
the house was spent scraping, finding some of the house's secrets, like
wallpaper underneath wallpaper. In the kitchen, a pattern of white-
gold trellises supported purple grapes as big and round as Ping-Pong

balls. When we painted the walls yellow, I thought of the bits of grape that remained underneath and imagined the vine popping through, the way some plants can tenaciously push through anything. The day of the big snow, when you had to shovel the walk and couldn't find your cap and asked me how to wind a towel so that it would stay on your head—you, in the white towel turban, like a crazy king of the snow. People liked the idea of our being together, leaving the city for the country. So many people visited, and the fireplace made all of them want to tell amazing stories: the child who happened to be standing on the right corner when the door of the ice-cream truck came open and hundreds of Popsicles cascaded out, the man standing on the beach, sand sparkling in the sun, one bit glinting more than the rest, stooping to find a diamond ring. Did they talk about amazing things because they thought we'd turn into one of them? Now I think they probably guessed it wouldn't work. It was as hopeless as giving a child a matched cup and saucer. Remember the night, out on the lawn, knee-deep in snow, chins pointed at the sky as the wind whirled down all that whiteness? It seemed that the world had been turned upside down, and we were looking into an enormous field of Queen Anne's lace. Later, headlights off, our car was the first to ride through the newly fallen snow. The world outside the car looked solarized.

You remember it differently. You remembered that the cold settled in stages, that a small curve of light was shaved from the moon night after night, until you were no longer surprised the sky was black, that the chipmunk ran to hide in the dark, not simply to a door that led to its escape. Our visitors told the same stories people always tell. One night, giving me a lesson in storytelling, you said, "Any life will seem dramatic if you omit mention of most of it."

This, then, for drama: I drove back to that house not long ago. It was April, and Allen had died. In spite of all the visitors, Allen, next door, had been the good friend in bad times. I sat with his wife in their living room, looking out the glass doors to the backyard, and there was Allen's pool, still covered with black plastic that had been stretched across it for winter. It had rained, and as the rain fell, the cover collected more and more water until it finally spilled onto the concrete. When I left that day, I drove past what had been our house.

Three or four crocus were blooming in the front—just a few dots of white, no field of snow. I felt embarrassed for them. They couldn't compete.

This is a story, told the way you say stories should be told: Somebody grew up, fell in love, and spent a winter with her lover in the country. This, of course, is the barest outline, and futile to discuss. It's as pointless as throwing birdseed on the ground while the snow still falls fast. Who expects small things to survive when even the largest get lost? People forget years and remember moments. Seconds and symbols are left to sum things up: the black shroud over the pool. Love, in its shortest form, becomes a word. What I remember about all that time is one winter. The snow. Even now, saying "snow," my lips move so that they kiss the air.

No mention has been made of the snowplow that seemed always to be there, scraping snow off our narrow road—an artery cleared, though neither of us could have said where the heart was.

—ANN BEATTIE, "Snow"

Frostbite is for losers.

—VINCE LOMBARDI

> After the officers course at Ft. Benning,
> a stint as training officer
> running various rifle ranges,
> then a chemical warfare course
> at McClellan, just outside
> Anniston, Alabama where
> the orientation officer says
> no one should be after dark—
> *Don't even stray off the main highway*
> he says, *It's not a good place to be,*
> I get orders, a unit waiting
> to be deployed to Viet Nam,
> also not a good place to be,
> December '67, and like they always said—
> motto of the Army—*Hurry up and wait:*

we just sat there, more training,
and plenty of time to think about it,
so we got together and watched football,
lieutenants newly commissioned
including a grizzled former E-6 from Texas
with enough enlisted time for half a career,
part Indian, a Cowboys fan, and me
from Wisconsin—and the title game
was tight, thirteen below, wind chill
minus seventy, the field brittle as ice,
Dallas ahead by three
in the fourth, less than a minute
to go, ball on the one, and Donnie Anderson
can't move it a single inch against
the frozen Dallas line—two running plays
and nothing but time on the clock eaten up
until there's but sixteen seconds left
forcing Starr to call his final timeout:
he confers with Lombardi, says he can
sneak it past the goal though the coach
wants Mircein to punch it in;
Lombardi capitulates, says *Do it*
and let's get the hell out of here.
Starr, ever a master of deception,
calls Brown right, thirty-one wedge,
but the blocking assignments are the same
and Starr keeps the ball instead
of handing off to Mircein:
on the snap Kramer slams Jethro Pugh,
jams him outside just enough for Starr
to run through the gap, fall and
stretch across the goal: the Texan,
the Indian lieutenant, is crushed
like a beer can under a tank,
and I almost feel sorry for him:
as Tom Landry said after the game,

It was a dumb call
in a world filled with dumb calls,
but elation knows little of mercy,
and next month there's Tet,
a hard year for everyone,
not an inch of ground gained
in a losing season, no matter how
you look at it, and he didn't do well
when he came back, cold sweats
in a Texas heat wave, knees stiff from shrapnel,
bone shattered like ice, an unacceptable loss
where *Winning isn't everything*
he keeps telling himself
as the temperature falls,
as the wind—it must be the wind—
sends an icy chill down his shivering back.

—DALE RITTERBUSCH, "Ice Bowl"

Entr'acte

Each day after the second surgery, I felt worse. I was informed that my body's wastes were rising and that my reconnected bowels misunderstood. I began to projectile vomit. The force of it all seemed dangerous, and the pain from the heaves made me sweat. The nausea caused by the constantly building bile felt impossible to face. I was weeping when the head nurse came into my room with equipment. She had a six-foot tube coiled in ice in a rectangular tray. The tube was transparent. Nurse sat me upright in a straight-backed chair in the center of the room and began feeding the tube up my right nostril. She'd asked if I had a preference, left? right?

The tube had been packed in the ice to stiffen—but not too stiff—because, as Nurse pointed out, it had to make the turn to descend by way of the throat to the stomach. The tube being force-wormed down my throat worsened all. Nurse kept saying, "Help me, Mr. Anderson." She had me nipping at water from a mini-cup. In theory, this sacramental nipping was to help me to not resist the tube. In time, the tube was installed by way of my left nostril (Nurse abandoning the attempt through the right), down my throat, to and through the cardia, into the well of my stomach. I wept throughout the procedure.

—DONALD ANDERSON, "Luck"

The crushing power of moving pack ice is not a great threat to people traveling with dogs or on foot. They can usually move nimbly enough over its surface. To be at its mercy in a boat or small ship, however, is to know an exhausting, nerve-wracking vulnerability. In May 1814, with his whaling ship beset off the east coast of Greenland, William Scoresby set out on foot to reconnoiter the final mile of maneuvering that he hoped would set him free. Like many men caught in such circumstances, Scoresby was terrified. But he was mesmerized as well by the ice, by it sheer power, its daunting scale, the inexorability of its movement. The sound of its constant adjustment before the wind was like "complicated machinery, or distant thunder," he wrote. Even as he sought a way out, he marveled at the way it distracted him. He lost the sense of plight that spurred him, the pleading whining that came from his ship's pinched hull; he became a mere "careless spectator." It was as though he were walking over the back of some enormous and methodical beast.

—BARRY LOPEZ, *Arctic Dreams*

With the utmost caution I picked my way through the sparkling bergs, and after an hour or two of this nerve-trying work, when I was perhaps less than halfway across and dreading the loss of the frail canoe which would include the loss of myself, I came to a pack of very large bergs which loomed threateningly, offering no visible thoroughfare. Paddling and pushing to right and left, I at last discovered a sheer-walled opening about four feet wide and perhaps two hundred feet long, formed apparently by the splitting of a huge iceberg. I hesitated to enter this passage, fearing that the slightest change in the tide-current might close it, but ventured nevertheless, judging the dangers ahead might not be greater than those I had already passed. When I had got about a third of the way in, I suddenly discovered that the smooth-walled ice-lane was growing narrower, and with desperate haste backed out. Just as the bow of the canoe cleared the sheer walls they came together with a growling crunch. Terror-stricken, I turned back, and in an anxious hour or two gladly reached the rock-bound shore that had at first repelled me, determined to stay on guard all night in the canoe or find some place where with the strength that in a fight for life I could drag it up the boulder wall beyond ice danger. This at last was happily done about midnight, and with no thought of sleep I went to bed rejoicing.

—JOHN MUIR, *Travels in Alaska*

About the middle of the afternoon we were directly opposite a noble group of glaciers some ten in number, flowing from a chain of crater-like snow fountains, guarded around their summits and well down their sides by jagged peaks and cols and curving mural ridges. From each of the larger clusters of fountains, a wide, sheer-walled cañon opens down to the sea. Three of the trunk glaciers descend to within a few feet of the sea level. The largest of the three, probably about fifteen miles long, terminates in a magnificent valley like Yosemite, in an imposing wall of ice about two miles long, and from three to five hundred feet high, forming a barrier across the valley from wall to wall. It was to this glacier that the ships of the Alaska Ice Company resorted for the ice they carried to San Francisco and the Sandwich Islands, and, I believe, also to China and Japan. To load, they had only to sail up the fiord within a short distance of the front in the terminal moraine.

—Ibid.

Cold Facts
1880. The first totally successful shipment of frozen beef and mutton from Australia to England arrives in early February as the S. S. *Strahleven* docks with four hundred carcasses. The meat sells in

London at 5.5 shillings per pound and is soon followed by the first cargo of frozen mutton and lamb from New Zealand.

US ice shipments to tropical ports reach a high of 890,364 tons carried by 1,735 ships, up from 146,000 tons aboard 363 ships in 1856. On long voyages 40 percent of the cargo may melt but the remaining ice sells for as much as $56 per ton.

1887. Western Cold Storage Co. in Chicago installs ice-making machines, but cold stores in most places continue to be cooled by harvested ice cut from lakes and hauled ashore by teams of horses.

1889. A US ice shortage caused by an extraordinarily mild winter gives impetus to the development of ice-making plants. By year's end the country has more than 200 ice plants, up from 35 in 1879.
—JAMES TRAGER, *The People's Chronology*

In the shadow of America's great inventors—Edison, Ford and Bell, to name a few—stands an unheralded giant: Clarence Birdseye, the father of the modern "fresh frozen" pea. Wander any supermarket and you'll find Birdseye's legacy, neatly wrapped packages containing every imaginable type of fast-frozen fruit, vegetable and meat, their natural colors, textures and flavors miraculously preserved by Birdseye's pioneering method.

Commonplace now, fast-frozen food, first marketed in 1930 under the label Bird's Eye Frosted Foods, was initially hailed as a marvel of science. For the first time, June sweet peas and summer blueberries could be savored, in close-to-fresh form, in the dead of winter. By the mid-1940s, Americans were eating over 800 million pounds of fast-frozen food a year.
—ABIGAIL MEISEL, "In From the Cold"

[Birdseye] came of age just as Theodore Roosevelt, America's cowboy-scholar-naturalist, was ascending in politics. Like Roosevelt, Birdseye fell sway to the romanticism of the American West and as a young man lit out for Arizona, New Mexico and Montana. He embraced physical challenges, surviving in Labrador for months at a stretch, enduring the isolation of arctic winters and keeping company with rugged adventurers. It was during a stay in this far northern region that Birdseye made his first observations about freezing and crystallization, and found, to his surprise, that the frozen food in Labrador was "not unpleasant. . . ."

—Ibid.

Before they set up drilling rigs in waters off the coast of Newfoundland and Alaska—commonly called "ice-infested" waters, as if the ice were rats or insects—they needed to know more about what would happen if ice hit a rig or if a tanker hit ice. The island was a 3,000-year-old, five-mile-long slab of mostly freshwater ice which had broken from an ice shelf off Ellesmere island eight years earlier and had been drifting in the company of sea ice ever since. By the time the Canadians set up a research camp on it, it was fewer than 600 miles from the North Pole. Except for a few low peaks sticking up off Ellef Ringnes, a rocky island in the distance named for a Norwegian beer manufacturer who supported early Arctic expeditions, the landscape surrounding the camp—the ice island, the sea ice attached to it—was more or less flat in all directions. And except for those low peaks and a few dark nuggets lying on top of the snow cover, clues that Arctic foxes had been coming around, the landscape was all white under 24-hour-a-day sunlight.

The tests were meant to mimic a sudden collision between ice and a metal rig or tanker, so that researchers would have an idea what damage the ice could do to the metal and the metal to the ice. After sawing a ten-foot-deep, 300-foot-long trench in the ice, engineers lowered into its glowing blue-green interior what was essentially a giant hammer (an indenter) with a metal plate about the thickness of a ship's hull on its face and wires running from computers to the sensors on the plate. Somebody flipped a switch, and with a shudder and a restrained crunching noise, the indenter hit one wall of the trench with a million pounds per square inch of force, equal to the weight of a thousand small cars sitting on top of each other, one engineer said. When the indenter was pulled away from the wall minutes later, the metal plate had a dent in it the size of a saucer and the wall had a circle of crushed white ice on it the size of the plate. Running through the white, however, was a large swirl of the same blue-green as the surrounding wall, indication that the ice there had not been crushed; in outline it looked rather like a map of Newfoundland.

—MARIANA GOSNELL, *Ice: The Nature, the History, and the Uses of an Astonishing Substance*

The Earth, the fabled water planet, is also an ice planet. More than 10 percent of the terrestrial Earth now lies under ice, with another 14 percent affected by periglacial environments and permafrost. Some 7 percent of the world is covered by sea ice, and at any minute nearly 25 percent of the world ocean is affected by ice, especially icebergs. The vast proportion of the bergs inhabit the Southern Ocean, corralled by the Antarctic convergence. Of the Earth's cryosphere, 99 percent is glacial ice, and 96 percent of that—over 60 percent of the world's freshwater reserves—is in Antarctica. Within past geologic eras, the proportion of ice on the Earth has grown enormously. During the last glaciation in the Pleistocene, ice extended over 30 percent of the planet's land surface and affected 50 percent of the world ocean. The immensity of the ice sheet even today is sufficient to deform the entire planet, so depressing the south polar region as to make the globe slightly pear-shaped.

—STEPHEN J. PYNE, *The Ice: A Journey to Antarctica*

And I—could I stand by
And see You—freeze—
Without my Right of Frost—
Death's privilege?
—EMILY DICKINSON, from "I Cannot Live Without You"

We found a number of carcases of the Buffaloe lying along shore, which had been drowned by falling through the ice in winter and lodged on shore by the high water when the river broke up about the first of this month.

—MERIWETHER LEWIS, *Journal*, April 13, 1805

A solemn, unsmiling, sanctimonious old iceberg that looked like he
was waiting for a vacancy in the Trinity.

—MARK TWAIN, letter to the *Alta California*

Most permafrost is old, very old. Since ice is a pretty good insulator,
the more permafrost there is for the latent heat of freezing to have to
pass through before reaching cold air, the more delayed the addition
of a new layer of ice at the bottom will be. Thus it takes a very long
time for even very cold air to create a very thick slab of permafrost.
Most of the permafrost in the world has been around not just for two
summers but for thousands of summers, some of it for hundreds of
thousands of summers. In Alaska and the Yukon there's permafrost
believed to be 2 to 2 ½ million years old.

—MARIANA GOSNELL, *Ice: The Nature, the History, and the Uses of
an Astonishing Substance.*

> The city moves slow as an oak branch
> under the burden of snow.
> Sheets of ice driftwood on the Hudson
> and waterpaths churn in the wake of ships
> I've seen many times from windows
> of high rises. Now a street musician
> has rolled a piano into the square,
> and though his hands must be numb
> all afternoon he plays variations of Bach
> and sometimes a song from the 80s.
> Last night was the first snowfall
> I've seen all winter, the white wash
> an alteration of the landscape.
> I'm leaving a trail through the square
> because someone should know
> I've been here, in case I don't return.

—LAREN MCCLUNG, from "Dog Weather"

In August of 1930, a Norwegian sloop, the *Bratvaag,* sailing in the
Arctic Ocean, stopped at a remote island called White Island. The

Bratvaag was partly on a scientific mission, led by a geologist named Dr. Gunnar Horn, and partly out sealing. On the second day, the sealers followed some walruses around a point of land. A few hours later, they returned with a book, which was sodden and heavy, and had its pages stuck together. The book was a diary, and on the first page someone had written in pencil, "The Sledge Journey, 1897."

Horn rode to shore with the *Bratvaag*'s captain, who said that two sealers dressing walruses had grown thirsty and gone looking for water. By a stream, Horn wrote they found "an aluminum lid, which they picked up with astonishment," since White Island was so isolated that almost no one had ever been there. Continuing, they saw something dark protruding from a snowdrift—an edge of a canvas boat. The boat was filled with ice, but within it could be seen a number of books, two shotguns, some clothes and aluminum boxes, a brass boathook, and a surveyor's tool called a theodolite. Several of the objects had been stamped with the phrase "Andrée's Pol. Exp. 1896." Near the boat was a body. It was leaning against a rock, with its legs extended, and it was frozen. On its feet were boots, partly covered by snow. Very little but bones remained of the torso and arms. The head was missing, and clothes were scattered around, leading Horn to conclude that bears had disturbed the remains.

He and the others carefully opened the jacket the corpse was wearing, and when they saw a large monogram A they knew whom they were looking at—S. A. Andrée, the Swede who, thirty-three years earlier had ascended with two companions in a hydrogen balloon to discover the North Pole.

—ALEC WILKINSON, *The Ice Balloon*

Antarctica, containing 90 percent of the world's ice (and almost 70 percent of its fresh water), is covered with ice to an average of 7,000 feet thick. If all of Antarctic's ice melted, sea levels around the world would rise some 200 feet. But the average temperature in Antarctica is -35°F, so the ice there is in little danger of melting.

At the North Pole, the ice is not nearly as thick as at the South Pole and floats on the Arctic Ocean. If this northern ice melted, sea levels would not be affected.

However, the amount of ice covering Greenland, if melted, would cause a rise of 20 feet to the oceans. Because Greenland is closer to the equator than Antarctica, its latent and potential temperatures are higher.

> —"If the polar ice caps melted, how much would the oceans rise?," Howstuffworks.com, http://science.howstuffworks.com/ environmental/earth/geophysics/question473.htm

Officially the Ice Age is still on, although we're in an inter-glacial hiatus, or an interstade, as geologists name it. The most recent advance ended only 10,000 to 12,000 years ago; the next, if the cycles are maintained, will return 50,000 to 100,000 years from now.

> —NEIL MATHISON, "Ice"

I encountered an art installation on the entry steps of a museum in Utah. It was July, and randomly placed on the granite steps were blocks of ice that were melting. Piled atop the blocks were dead tree leaves, oak and maple. I was sure that the drenched leaves would, in time, prove fatal . . .

In bitter cold butter got hard as stone, meat needed to be split like cordwood, and mercury could be fired from a gun. In parts of Siberia milk was sold in pieces. One explorer's account describes Eskimos using cold for revenge on wolves that had attacked them. Into the ice the Eskimos set several sharp knives with their blades upright; then they covered the blades with blood. The wolves licked the blades and cut their tongues, but felt nothing because of the cold. Their own blood on the blades kept them licking "until their tongues were so scarified that death was inevitable."

> —ALEC WILKINSON, The Ice Balloon

2012. NASA's Messenger spacecraft has discovered evidence that the planet Mercury has enough ice on its surface to encase Washington, D.C., in a block two and a half miles deep. "For more than 20 years the jury has been deliberating on whether the planet closest to the Sun hosts abundant water ice in its permanently shadowed polar regions," writes Sean Solomon of the Columbia University's Lamont-Doherty Earth

Observatory, the principal investigator of the *Messenger* mission. The spacecraft "has now supplied a unanimous affirmative verdict."

—"NASA: There's enough ice on Merucry to encase Washington, D.C.," Yahoo! News, http://news.yahoo.com/blogs/sideshow/nasa-says-enough-ice-mercury-encase-washington-dc-194415297.html

Now, at the beginning of the twenty-first century, glaciers cover close to 10 percent of Earth's land surface, mostly in two giant ice sheets over Greenland and Antarctica. Greenland's is nearly as large as Mexico, and in places almost two miles thick; Antarctica's is about one and a half times the size of the continental United States and up to nearly two and a half miles thick. Together, these two areas of ice contain enough water to raise global sea levels something like 230 feet. Ice fields and caps remain in Iceland, Alaska, Canada, and Patagonia. Glaciers still exist in most of the world's high mountain ranges, where they provide water to millions of people: nearly 20 percent of the world's people depend on ice and snow melt for their water. Nearly all—something like 98 percent of all mountain glaciers, and more than 90 percent of all glaciers—have been melting at alarming rates.

—GERALD DELAHUNTY, from *The Face of the Earth*

In certain rare conditions of wind and sunlight, glacial ice evaporates immediately, without passing through the liquid stage. This is called sublimation, a more refined form of melting.

The phenomenon is often accompanied by a rhythmic crackling sound, as if invisible feet were stepping across the ice.

—THOMAS WHARTON, *Icefields*

No matter their size when they begin life, once adrift, icebergs' disintegration is swift. The processes of melt and erosion are so relentless that once an iceberg breaks free of the pack ice, it fragments very quickly. When it leaves its only home and sets out on its reckless quest, its days are numbered. Within two months at sea all but the largest icebergs have disappeared.

—JEAN MCNEIL, *Ice Diaries: An Antarctic Memoir*

The air was dry, as if we were inhaling paper. Meltwater rivulets coursed alongside the paths but it lacked the dewy, ionized smell of water. There were no smells of soil; no trees, nor grass nor flowers. I did perceive a smell that I would never experience again, except to a muted degree in Greenland, but if I were to smell it now, I would recognize it immediately. When asked what the moon smelled like, astronaut Buzz Aldrin, the second man to walk on its surface, said, "Two stones rubbed together." That's the smell of the Antarctic—flint.

—Ibid.

White men always think of ice as frozen water, but Eskimos think of water as melted ice . . .

—JERRY KOBALENKO, *The Horizontal Everest*

"Hokkaido," the Pilot said. "The ice up there is worse in the summer. The shelves break up, currents churn it. It's the ice you don't see, that's what gets you.

The Captain spoke. Shirtless, you could see all his Russian tattoos. They looked heavy in the sideways light, as if they were what had pulled his skin loose. "The winters up there," he said, "everything freezes. The piss in your prick and the fish gore in your beard. You try to set a knife down and you can't let go of it. Once, we were on the cutting floor when the ship hit a growler. It shook the whole boat, knocked us down into the guts. From the floor, we watched that ice roll down the side of the ship, knuckling big dents in the hull."

—ADAM JOHNSON, *The Orphan Master's Son*

Look at a map, and it's easy to see why the Greenland ice sheet is so vulnerable: Its southern end is no farther north than ice-free Anchorage or Stockholm. Greenland's ice is a relic of the last ice age, surviving only because it is massive enough to make its own climate. The island's brilliant, perpetually snow-covered interior reflects light and heat. Its elevation adds to the chill, and its bulk fends off warm weather systems from farther south. As the ice sheet shrinks, all these defenses will weaken.

—"The Big Thaw," National Geographic, http://ngm. nationalgeographic.com/2007/06/big-thaw/big-thaw-text/6

Over the Alps, packed deep in hay and snow,
The Romans hauled their oysters south to Rome:
I saw damp panniers disgorge
The frond-lipped, brine stung
Glut of privilege.

—SEAMUS HEANEY, from "Oysters"

Escallops Of Penguin Breasts
Penguin Breasts as required
Reconstituted onion
Some fairly thick batter
Flour
Salt and pepper to taste
Cut the breasts into thin slices and soak in milk for about 2 hours.
Dry, season, and flour them well on both sides; have ready some deep
frying fat. When just smoking hot, dip the pieces in the batter with
the onion mixed into it and fry each piece to a nice golden brown. For
a sauce, turn the contents of a tin of mushroom soup into a saucepan
and heat but do not boil. When hot, pour over the meat and serve
with fried potatoes and peas.

—JASON C. ANTHONY, HOOSH: Roast Penguin, Scurvy
Day, and Other Stories of Antarctic Cuisine

Entr'acte

Our mucking machine, like everyone's, powered by hydraulics, was
a scoop—a steel bucket on cables and pulleys rock-bolted into the
face of the stope. When the cable snapped, my partner and I dragged
it, like road kill, to the tracks where we laid it on one of the ore car
steel rails. Here, he meant to trim the frayed ends in order to re-splice
them. My partner knelt by the rail and arranged the heavy snake. I
handed him the axe.

I was asking questions and, in the process of answering me and trimming the cable, the man lopped off his left index finger. It was a powerful swing and a clean cut, just shy of the knuckle, toward the wrist. The man removed his glove to stanch the bloody pump. He didn't shriek when the axe fell or when he observed the result. He wrapped the wound with shirt cloth he'd torn away with one swift move with his good right hand. Shit, he murmured, then rose to stand in his half shirt. He could have been a statue, a bronze man in a helmet, a hero, an actor in a movie starring Victor Mature. The miner's reaction made me think of Ulysses or Samson.

I stooped to retrieve the digit. It was still in its little glove case, like a gift jackknife. We headed for the station, where a massive box constructed of lagging (rough cut 2×8s) housed coils of pipe through which water flowed for drinking, for drilling, and for quelling dust. Each morning, on each level of the mine (we were on the 5200 level, a mile deep) block ice was shipped down and dumped into these boxes to chill the water.

I had carried the axe, and chopped on a block for ice to fill my hardhat. I dumped the finger from the glove, then covered it with ice.

At the lighted station, we rang for the cage. My father, in his role as safety engineer, arrived with the shift boss. He transferred the finger and ice to his hat. My partner, my father, the shift boss, and the iced bone and flesh were hoisted up. My father congratulated me for thinking of ice. He called me *Donnie* in front of the men.

I reattached the lamp to my hardhat and hoofed it back to the stope and the cable. There were ice chunks caught in the hardhat webbing. I stuck two pieces in my mouth. I ate lunch, then looked at my watch and waited. Temperatures at 5200 feet were generally in the 90s. I was sweating when I saw light approaching. I guessed it was my father coming to check on me, but it was my partner. He had been to St. James Hospital where his finger had been reattached. My partner walked me through resplicing the snapped cable, and we went back to work with me operating the mucking machine. In the middle of it all, I shot my partner a look. Didn't want to miss the shift, he said. He held the ungloved hand in front, as if to know where it was. Its wrapping shone a frosted white in our lamplight.

—DONALD ANDERSON, *Gathering Noise from my Life*

Rods and cones
at the ends of fingers held up to the sun.
Snakes work in their roots of sleep
and dirt shows through
a huge fingerprint in the field.
Hiss of crystals breaking
as sun renders out the fat in snow
to run and be drawn to the sky again.

—ROBERT MORGAN, "Thaw"

James Balog

—see *Chasing Ice*

The chicken was going to be a present for a man who lived in the country and owned ducks, geese, and a swan. One thing I knew about this man was that he liked his birds the way some people like dogs and cats, and he probably wouldn't eat them. I was trying to picture the chicken in his new home when I crossed a bridge over the Susquehanna and encountered the silence of black ice. The tires lost their hiss, the chicken shut up, and about fifty yards after I hit the ice, I hit a Tioga County Sheriff's Department car. The car was parked on the road berm, just beyond the bridge, and inside the car a sheriff was radioing for a tow truck, as if he knew I was coming and that when I got there, our two cars were going to need help.

My car did a kind of simple dance step down the highway on its way to meet the sheriff's car. It threw its hips to the left, it threw its hips to the right, left, right, left, right, then turned and slid, as if it were making a rock-and-roll move toward the arms of a partner.

—STEPHANIE VAUGHN, "We're On TV in the Universe"

Before the impact, when my car was still grace on ice, when my car was no longer in touch with the planet but now sliding above a thin layer of air and water, four thousand pounds of chrome and steel, bronze metallic paint, power steering, power brakes, AC, AM/FM, good tires, fine upholstery, all the things you like to see in an ad when you're looking for a big, used American car, when it was gliding through that galaxy of flashing lights, on its way through Andromeda, Sirius, and the Crab Nebula, it crossed my mind that surely it was against the laws of physics to hit a patrol car. If you were sliding above ice, you might hit a regular car, or a pole, or a fence, or an asteroid, but you could not hit the car of a man with a badge, a gun, bulletproof windows, citation forms in his pocket, handcuffs, the power to arrest you, a man working hard on a bad night.

—Ibid.

Like the best of worst things,
I never saw it coming.
But it saw me.

My bifocals,
the heavy tread of my boots,
my still-quick reflexes,
it just laughed at.

The fluff of a feathery morning,
the melt of midday,
the rainbow puddles of evening all
devolved into this cold black hole
of beauty that, like Rilke's angels,
could clasp and kill.

Perhaps my tweed cap,
from Granddaddy's village in Wales,
saved me, cushioned my head
when it slapped back to slap
the black pavement where cars
were parked all day,
their candy colors warming,
dripping, and drooling in weak sun.

I crawled on my hand and knees,
groping in the snow for specs,

new cell phone, hat,
calling out what I thought
I never would:
help.

And then, oh, what have I done?
over and over, since my wife
lay stroked and helpless

in a hospital room upstairs.

All night I drank gin,
called friends to wake me for
what could be my final hour,
staring at the mirror,
at the dark disks of pupils
to see if they would dilate,
or deepen the pool
I might fall into
If something brittle didn't
Bear me up from that blackness

that has no reflection
and no bottom.

—WILLIAM GREENWAY, "Black Ice"

Entr'acte

I had torn off my side-view mirror on a concrete pillar in the underground garage where I work. On a blizzard day—eighteen inches of snow and high winds—schools and government offices had closed, and I was home from work. I called WRECKMASTERS: *Wreckmasters—Jimbo here.* "So you're there," I said. "I wondered." *Our kind of weather,* Jimbo said.

Frederic Tudor cut the ice from New England ponds . . . and packed it in ships and sent it to hot countries like Australia. . . . Maybe in the future, Americans will pack our ships, like Tudor did, with ice from our many hotels and general stores and personal ice makers. Maybe, instead sending our ice to the hotter countries, we'll send it to the previously colder ones. I can see these ships, sailing toward the Arctic Circle, loaded down with rattling, telltale cubes, our country sacrificing its right to cold sodas and perfect cocktails in order to save the planet. When the ships dock, Americans will unload their melting cargo, each carrying a plastic ice bucket on which a lid can be placed, and the contents delivered, like cold ashes in an urn, to its final destination on the glacier.

—HEIDI JULAVITS, "American Exceptionalism on Ice"

Try to imagine a world without ice. No ice to chill our gin and tonics. No ice to give us Peggy Fleming. No Yosemite Valley. No Half Dome. No Lake Superior. No Apolo Ohno. No ice hockey. No Red Wings. No Bruins. No glacial crevasses and no moraines. No eskars or drumlins or wandering erratic rocks. No snowflakes or hail or frost-whitened mornings. No snowmobiles and no skis. No horse-drawn sleighs. No island of Manhattan. No Puget Sound.

—NEIL MATHISON, "Ice"

Earth has broken monthly heat records 25 times since the year 2000, but hasn't broken a monthly cold record since 1916.

—*THE DENVER POST*, July 21, 2015

Researchers knew California's drought was already a record breaker when they set out to find its exact place in history, but they were surprised by what they discovered: It has been 500 years since what is now the Golden State has been this dry.

California is in the fourth year of a severe drought with temperatures so high and precipitation so low that rain and snow evaporate almost as soon as they hit the ground.

A research paper released Monday said an analysis of blue oak tree rings in the state's Central Valley showed that weather conditions haven't been this dire since the 1500s.

That was about the time when European explorers landed in what became San Diego, when Columbus set off on a final voyage in the Caribbean, when King Henry VIII was alive. . . . The study was published in the journal *Nature Climate Change.*

—DARRYL FEARS of *The Washington Post*, quoted in *The Denver Post*, September 15, 2015

Many years later, as he faced the firing squad, Colonel Aureliano Buendía was to remember that distant afternoon when his father took him to discover ice.

—GABRIEL GARCÍA MÁRQUEZ, *One Hundred Years of Solitude*

Cold Facts

200 years—The predicted time for the collapse (directly due to global warming) of a cluster of Antarctica glaciers

4 feet—The predicted rise in sea level after the collapse

80 percent—The share of the world's fresh water that is in Antarctica

2.5 miles—The measurement of the ice sheet at its thickest point

8.7 million—In square miles, the size of Antarctica, the world's largest ice sheet

40 years—The length of time during which this data was gathered and analyzed

—see http://www.antarcticglaciers.org/

Some say the world will end in fire,
Some say in ice.

—ROBERT FROST, from "Fire and Ice"

A report released by the Centers for Disease Control and Prevention analyzed U.S. deaths attributed to the cold, the heat, storms, floods and lightning. The weather kills at least 2,000 Americans each year and nearly two-thirds of the deaths are from the cold, according to the report.

—*THE DENVER POST*, July 30, 2014

Cold Facts

... [for] every person who drives a car one thousand miles or takes a round-trip flight from New York to London, about thirty-two square feet of sea ice vanishes from the Arctic.

... for every metric ton of carbon dioxide emitted into the atmosphere, about three square meters of Arctic sea ice melts.

The average American emits more than sixteen metric tons of carbon each year ... melting enough ice to cover the floor plan of a fifty-five-square-foot apartment.

At the current rate of emissions, the Arctic could be free of ice by mid-century ...

"The Arctic is not like Las Vegas. What happens in the Arctic doesn't necessarily stay in the Arctic."
—KARL RITTER, *The Associated Press*, quoted in
The Denver Post, Friday, November 4, 2016

In fact, Donald J. Trump did say that climate change was a hoax invented by the Chinese ...

> Just looking at the scratches on
> this rock, all cut in the same direction,
> you'd never guess the little scores
> were tracks of mighty glaciers.
> What seems the grooves of rake or comb
> are traces of a crushing dome
> that passed this way first going south
> and then with thaw retreating north,
> the work of far millennium
> writ small as epitaph for tomb.
> —ROBERT MORGAN, "Toothmarks"

What are the chances that the Trump family recycles?

A study this year found Alaska's glaciers are losing 75 billion metric tons of ice per year. Alaska's glaciers comprise 11 percent of the world's total, but contributing 25 percent of the losses. Glaciers aside, when the permafrost goes, old plants thawing from the frozen ground will decompose to release tons and tons (multiplied by some number no one wants to know) of carbon. Eighty percent of Alaska is underlain by at least some permafrost.
— THE DENVER POST, September 1, 2015

As the iceberg shears off the submarine periscope, the noise
is less groan, more wild animal shriek. "Trust me," said the captain

piloting toward gunfire to see what the Russians are up to these days.
The sea ice resembles a cracked white lung steadily swelling

then sinking as high tide fades away. Already birds and
barnacles and butterflies are shifting their habitats poleward,

the eelgrass and jellyfish will be fine, but the basements
of coastal cities will begin to flood, an inch at a time.

The polar bear at the zoo makes the child start to cry:
why doesn't he move? Animals who cannot acclimate

to shifting conditions engender scientific argument
over what breaks first: the heart or the brain. In the heart

of the Arctic, underwater microphones listen for enemy traffic.
The noise made by a million barnacle larvae swimming north

is less hiss or whisper, more betrayed stare. When rations ran low,
polar explorers ate one less biscuit. When biscuits ran out,

the horses were first to be shot. In another sixty thousand years
the mouth of the Beardmore Glacier will spit out their bones.
— JYNNE DILLING MARTIN, "What Breaks First"

Iceland

The Solheimajokull glacier shriveled by about 2,050 feet between 2007 and 2015.

Alaska

The forward edge of the Mendenhall glacier outside of Juneau has receded about 1,800 feet between 2007 and 2015.

Switzerland

The Stein glacier has shrunk about 1,800 feet between 2006 and 2015.

Switzerland

The Trift glacier retreated nearly three-quarters of a mile between 2006 and 2015.

Peru

Ohio State ice scientist Lonnie Thompson has visited the Qori Kalis glacier since 1974. Between 1978 and 2016, it has shriveled 3,740 feet. Thompson described his regular expeditions to the Peruvian glacier "like visiting a terminally ill family member."

—see *THE DENVER POST*, April 5, 2017

The ice sheet is a holdover from the last ice age, when mile-high glaciers extended not just across Greenland but over vast stretches of the Northern Hemisphere. In most places—Canada, New England, the upper Midwest, Scandinavia—the ice melted away about ten thousand years ago. In Greenland it has—so far, at least—persisted. At the top of the sheet there's airy snow, known as firn, that fell last year and the year before and the year before that. Buried beneath is snow that fell when Washington crossed the Delaware and, beneath that, snow from when Hannibal crossed the Alps. The deepest layers, which were laid down long before recorded history, are under enormous pressure, and the firn is compressed into ice. At the very bottom there's snow that fell before the beginning of the last ice age, a hundred and fifteen thousand years ago.

The ice sheet is so big—at its center, it's two miles high—that it creates its own weather. Its mass is so great that it deforms the earth, pushing the bedrock several thousand feet into the mantle. Its gravitational tug affects the distribution of the oceans.

In recent years, as global temperatures have risen, the ice sheet has awoken from its postglacial slumber. Melt streams like the Rio Behar have always formed on the ice; they now appear at higher and higher elevations, earlier and earlier in the spring. This year's melt season began so freakishly early, in April, that when the data started to come in, many scientists couldn't believe it. "I had to go check my instruments," one told me. In 2012, melt was recorded at the very top of the ice sheet. The pace of change has surprised even the modellers. Just in the past four years, more than a trillion tons of ice have been lost. This is four hundred million Olympic swimming pools' worth of water, or enough to fill a single pool the size of New York State to a depth of twenty-three feet.

—ELIZABETH KOLBERT, "Letter From Greenland: A Song of Ice"

I first visited the Greenland ice sheet in the summer of 2001. At that time, vivid illustrations of climate change were hard to come by. Now they're everywhere—in the flooded streets of Florida and South Carolina, in the beetle-infested forests of Colorado and Montana, in the too warm waters of the Mid-Atlantic and the Great Lakes and the Gulf of Mexico, in the mounds of dead mussels that washed up this summer on the coast of Long Island and the piles of dead fish that coated the banks of the Yellowstone River.

But the problem with global warming—and the reason it continues to resist illustration, even as the streets flood and the forests die and the mussels rot on the shores—is that experience is an inadequate guide to what's going on. The climate operates on a time delay. When carbon dioxide is added to the atmosphere, it takes decades—in a technical sense, millennia—for the earth to equilibrate. This summer's fish kill was a product of warming that had become inevitable twenty or thirty years ago, and the warming that's being locked in today won't be fully felt until today's toddlers reach middle age. In effect, we are living in the climate of the past, but already we've determined the climate's future.

—Ibid.

The center of our universe—*our fat, lolling sun*—is unlikely to grant us a break, brightening, as she does, by 10 percent every billion years. Carbon emissions aside, we have something less than a billion years to figure something out . . .

> Snow is beginning to fall,
> huge wet flakes that burst from
> the darkness like parachutes
> and plunge past the streaming light
> and melt into the street. Freeze,
> die, says the veteran wind
> from the north, but he goes on
> with his work, the night, the snow,
> and was not speaking to me.
>
> —ROBERT MEZEY, from "Night on Clinton"

> His hindquarters must have fallen through
> the ice, and he could not pull himself back out
> and the incoming colder weather
> refroze the hole around him and he died,
> sinking some, only his broad horns
> holding his head and neck above the surface.
> Soon he must have been discovered by coyotes,
> who ate all they could.
>
> —ROBERT WRIGLEY, from "Elk"

> Although the coyotes may be back tonight,
> to dig their way from the horns' stumps
> for the ears, which I notice are still whole and upright,
> the left one turned slightly farther left,
> as though, with the last of his miraculous
> senses, he heard them coming over the ice.
>
> —Ibid.

Entr'acte

Bill Secours ran the power plant at the Air Force radar site in Alaska, the plant consisting of seven diesel engines, five of which were the minimum to keep the site heated and lighted. A Chicago son, Secours had worked in Alaska since his discharge at the end of World War Two, already some thirty years earlier by the time I met him. He was a big man who kept to himself. Half his scalp and one ear were gone. He spent most of his time in the power plant or in his room, which was situated next to mine in the pre-fab barracks arrangement where we (the officers and senior civilians) lived. Bill had a console record player on which he played what must have been every high-fidelity album ever submitted by Les Paul and Mary Ford (*Vaya Con Dios, How High the Moon, Mockin' Bird Hill*, et al.). New Year's Eve ("Tennessee" Travis one day gone) Secours showed in the break room to play dominoes with me, the commander, and the fire chief. I was a newly minted captain, the commander a lieutenant colonel, and the fire chief, a senior master sergeant. We were, all told, 115 men assigned to Indian Mountain Air Force Station situated in the Brooks Range, beside the high-flowing Indian River, electronically scanning the sky for the Russian Bear. Unlike the rest of us, Bill Secours wasn't serving a year-long involuntary tour; he'd been at Indian Mountain, voluntarily, for what seemed to the rest of us, forever. That night at dominoes I asked Bill if he'd ever taken a vacation to visit home. He grunted. A bit later (and drunk) I asked what'd happened to his head. *Grizzly*, he said. The regulation at the site was that you were not to move beyond two hundred yards of your dwellings without a weapon (of at least 350 magnum), something to stop a bear or moose. Bill had been armed when the grizzly attacked. He'd owned a Weatherby of the sort hunters cart to Africa to hunt rhinos or elephant. *Elephant*? I asked. *Yes*. It was the most I'd heard Secours talk, so I pressed. I asked if he'd tried to shoot the bear. *He had*, he said, *but before he could manage, the bear ate the thing. Ate it*? I said. *Yes*. He meant the Weatherby, not his head. I asked for advice to survive in the North. *Don't eat yellow snow*, he advised, serious as a bear attack.

Coda

Frost-*bite* acquires new meaning as the dermatologist zaps five precancerous growths on my pate with liquid nitrogen. When I flinch at the first spray from the stainless steel flask, the doctor admits, "I should have warned you." The flask seems oversized to me—like something you'd pump up to slaughter weeds in your yard. I sit still for the next four events, but feel as if roofing nails are being tacked into my skull. The treated areas feel curiously warm, though they are in fact flash-frozen-killed cells. I know that liquid nitrogen is cold, though am unprepared for its real number: -320.44°F. There are charts to show how long you can be exposed to temperature before the result of frostbite. A temperature of 0 degrees Fahrenheit, for instance, with a slight wind triggers frostbite in thirty minutes. Colder temperatures and stiffer wind reduce this window. If you are wet, all quickens. Which explains why aviators in the Arctic over water might not opt to bail out in case of flight failure. Nitrogen is 78 percent of the atmosphere. It is colorless, odorless, tasteless, and nontoxic. It boils at -320°F, is nonflammable, and will *not* support life. On vaporization, liquid nitrogen expands by a factor of almost seven hundred and may cause the explosion of a sealed container. Nitrogen can displace oxygen to cause asphyxiation. Neither I nor my dermatologist sport eye protection or insulated gloves. We do, though, sport closed-toe shoes and long pants. A *dewar* (named for its inventor, the Scotsman James Dewar) is a specialized vacuum flask that holds and can dispense liquid gas. In the nineteenth century, Dewar figured out that a vacuum between two flasks prevents the transfer of energy that occurs through conduction or convection. Furthermore, a silvered flask wall reflects, rather than absorbs, energy. Dewar did not bother to patent his ideas, and everything moved, unimpeded, to the insulated beverage container better known as Thermos™. Later, Dewar did manage to retain his patent for smokeless gunpowder (cordite) in a court case against Alfred Nobel, the creator of TNT. That positive aside, Dewar Liquid Nitrogen Flasks should, no doubt, be timely evaluated for safety, like elevators, wouldn't you think? The entire universe, it should be noted, is colder than liquid nitrogen. Before you consider deep space, realize that the average temperature of the universe is -452°F, some seven degrees from absolute zero wherein all molecular motion will cease. *Stay home.*

Acknowledgments

Appendix
A Selected Who's Who

Sonja Henie
Wayne Gretzky
Jan Ingemar Stenmark
Franz Klammer
Peggy Fleming
Sidney Crosby
Lindsey Vonn
Dorothy Hamill
Picabo Street
Jean-Claude Killy
Shaun White
Peter Prevc
Billy Kidd
Alberto Tomba
Bonnie Blair
Gordie Howe
Bode Miller
Dan Jansen
Matti Nykänen
Sara Takanashi
Freon /frëän/

Sources and Further Reading

Alexander, Caroline. *The Endurance: Shackleton's Legendary Antarctic Expedition*. New York: Knopf, 1998.

Amis, Martin. *Koba the Dread: Laughter and the Twenty Million*. New York: Vintage, 2002.

Anderson, Donald. "Fire Road." *Fire Road*. Iowa City: University of Iowa Press, 2001.

———. *Gathering Noise from My Life: A Camouflaged Memoir*. Iowa City: University of Iowa Press, 2012.

———. "Luck." In *Fire Road*. Iowa City: University of Iowa Press, 2001.

———. "Rock Salt." *EPOCH* 56, no. 2 (2007).

———. "Sao Paulo," unpublished.

———. "Scaling Ice." In *Fire Road*. Iowa City: University of Iowa Press, 2001.

———. "Stumps." *The North American Review* 289, no. 1 (2004).

Anthony, Jason C. *HOOSH: Roast Penguin, Scurvy Day, and Other Stories of Antarctic Cuisine*. Lincoln: University of Nebraska Press, 2012.

Atwood, Margaret. "Habitation." In *Selected Poems*. Toronto: Oxford University Press, 1976.

Baker, Nicholson. "Ice Storm." In *The Size of Thoughts*. New York: Vintage, 1997.

Bardach, Janusz, and Kathleen Gleeson. *Man Is Wolf to Man: Surviving the Gulag*. Berkeley: University of California Press, 1999.

Beattie, Ann. "Snow." In *Where You'll Find Me*. New York: Simon & Schuster, 1986.

Beevor, Antony. *Stalingrad: The Fateful Siege, 1942–1943*. New York: Viking, 1998.

Bickel, Lennard. *Mawson's Will: The Greatest Polar Survival Story Ever Written*. Hanover, NH: Steerforth, 2000.

Bragg, Rick. *All Over but the Shoutin'*. New York: Random House, 1997.

Braverman, Blair. *Welcome to the Goddamn Ice Cube: Chasing Fear and Finding Home in the Great White North*. New York: HarperCollins, 2016.

Busch, Benjamin. *Dust to Dust: A Memoir*. New York, HarperCollins, 2012.

Byrd, Richard. *Alone*, qtd. in Pyne.

Cecil, Richard. "One Hundredth Anniversary Edition of 'Birches.'" *River Styx* 89 (2013).

Chandler, David G. *The Campaigns of Napoleon*. New York: Macmillan, 1966.

Cheever, John. *Oh What a Paradise It Seems*. New York: Knopf, 1982.

Chew, Allen F. *White Death: The Epic of the Soviet-Finnish Winter War*. Lansing: Michigan State University Press, 2002.

Coleridge, Samuel Taylor. "The Rime of the Ancient Mariner." The Literature Network. http://www.online-literature.com/coleridge/646.

Couture, Pauline. *Ice: Beauty, Danger, History*. Toronto: McArthur & Company, 2004.

Craig, William. *Enemy at the Gates: The Battle for Stalingrad*. Old Saybrook, CT: Konecky & Konecky, 1973.

Delahunty, Gerald. *The Face of the Earth*. Edited by SueEllen Campbell. Berkeley: University of California Press, 2011.

Dickinson, Emily. *The Complete Poems of Emily Dickinson*. New York: Little Brown, 1960.

Dillard, Annie. "An Expedition to the Pole." In *Teaching a Stone to Talk: Expeditions and Encounters*. New York: Harper Collins, 1982.

Dubus, Andre III. *Townie*. New York: W. W. Norton, 2011.

Eire, Carlos. *Waiting for Snow in Havana*. New York: Free Press, 2003.

Emerson, Ralph Waldo. *The Essential Writings of Ralph Waldo Emerson*. New York: Modern Library, 2000.

Frost, Robert. *The Poetry of Robert Frost: The Collected Poems*, Complete and Unabridged. New York: Henry Holt, 1969.

———. "Robert Frost Quotes." Goodreads.com. http://www.goodreads.com/quotes/61934-like-a-piece-of-ice-on-a-hot-stove-the.

Gosnell, Mariana. *Ice: The Nature, the History, and the Uses of an Astonishing Substance*. New York: Knopf, 2006.

Greenway, William. "Black Ice." *River Styx* 87 (2012).

Haas, Robert, ed. *HAIKU: Versions of Basho, Buson, and Issa*. New York: Ecco, 1994.

Haggerty, John. "Tumbleweeds." In *War, Literature & the Arts: An International Journal of the Humanities* 21 (2009).

Halberstam, David. *The Coldest Winter: America and the Korean War*. New York: Hyperion, 2007.

Hansen, Ron. "Wickedness." In *Nebraska*. New York: The Atlantic Monthly Press, 1989.

Hastings, Max. *The Korean War*. New York: Simon and Schuster, 1987.

Hayden, Robert. "Those Winter Sundays." Poetry Foundation. https://www.poetryfoundation.org/resources/learning/core-poems/detail/46461.

Heaney, Seamus. *Opened Ground: Selected Poems, 1966–1996*. New York: Farrar, Straus and Giroux, 1998.

Heynen, Jim. "What Happened During the Ice Storm." In *The One-Room Schoolhouse: Stories About the Boys*. New York: Vintage, 1994.

Hochman, William. "Drive From Butte." In *Where We Are: The Montana Poets Anthology*, with an introductory note by James Wright, edited by Lex Runciman and Rick Robbins. Missoula, MT: CutBanks / SmokeRoot Press, 1978.

Hoover, Paul. *Winter (Mirror)*. Chicago: Flood Editions, 2002.

Howe, Nicholas. *Not Without Peril: 150 Years of Misadventure on the Presidential Range of New Hampshire.* Guilford, CT: Globe Pequor Press, 2009.

Humphreys, Helen. *The Frozen Thames.* Toronto: McClelland & Stewart Ltd., 2007.

Johnson, Adam. *The Orphan Master's Son.* New York: Random House, 2012.

Joyce, James. *Dubliners.* New York: Norton Critical Edition, 2006.

Julavits, Heidi. "American Exceptionalism on Ice." *New Yorker,* July 8, 2016.

Kobalenko, Jerry. *The Horizontal Everest: Extreme Journeys on Ellesmere Island.* Toronto: Penquin, 2002.

Kolbert, Elizabeth. "Letter From Greenland: A Song of Ice." *New Yorker,* October 24, 2016.

Kooser, Ted. "Late February." In *Sure Signs.* Pittsburgh: University of Pittsburgh Press, 1980.

Koschorrek, Gunter K. *Blood Red Snow: The Memoirs of a German Soldier on the Eastern Front.* Minneapolis: Zenith Press, 2005.

Krakauer, Jon. *Into the Wild.* New York: Doubleday, 1996.

———. *Into Thin Air: A Personal Account of the Mount Everest Disaster.* New York: Random House, 1997.

Kranes, David. "Hunt." In *Hunters in the Snow.* Salt Lake: University of Utah Press, 1979.

Lansing, Alfred. *Endurance: Shackleton's Incredible Voyage.* New York: Carroll & Graf, 1959.

Laskin, David. *The Children's Blizzard.* New York: HarperCollins, 2004.

Lewis, Meriwether. *Journal,* qtd. in Gosnell.

London, Jack. *Jack London: Novels and Stories.* New York: Library of America, 1982.

Lopez, Barry. *Arctic Dreams: Imagination and Desire in a Northern Landscape.* New York: Charles Scribner's Sons, 1986.

Lueders, Edward. *The Clam Lake Papers: A Winter in the North Woods.* New York: Harper & Row, 1977.

Martin, Jynne Dilling. *We Mammals in Hospitable Times.* Pittsburgh: Carnegie Mellon University Press, 2015.

Márquez, Gabriel García. *One Hundred Years of Solitude.* New York: Harper Perennial, 1970.

Mathison, Neil. "Ice." *The North American Review* 297, no. 1 (2012).

Matthews, William. *Rising and Falling.* New York: Houghton Mifflin Harcourt, 1979.

McClung, Laren. *Between Here and Monkey Mountain.* Rhinebeck, NY: The Sheep Meadow Press, 2012.

McGuane, Thomas. "Ice." In *Gallatin Canyon.* New York: Knopf, 2006.

McNeil, Jean. *Ice Diaries: An Antarctic Memoir.* Toronto: ECW Press, 2016.

Meisel, Abigail. *The New York Times Book Review,* July 8, 2012.

Melville, Herman. *White-Jacket.* New York: Library of America, 1983.

Mezey, Robert. *Collected Poems: 1952–1999.* Fayetteville: The University of Arkansas Press, 2000.

Morgan, Robert. "Immune." In *Terroir*. New York: Penguin, 2011.

———. "Thaw." In *Red Owl*. New York: W. W. Norton, 1972.

———. "Toothmarks." In *Dark Energy*. New York: Penguin, 2015.

Muir, John. *The Eight Wilderness Discovery Books*. Seattle: The Mountaineer's Books, 1992.

Munro, Alice. *The Progress of Love*. New York: Knopf, 1985.

Pinker, Stephen. *The Language Instinct*. New York: William Morrow & Company, 1994.

Pyne, Stephen J. *The Ice: A Journey to Antarctica*. Iowa City: University of Iowa Press, 1986.

Rarick, Ethan. *Desperate Passage: The Donner Party's Perilous Journey West*. New York: Oxford University Press, 2008.

Read, Piers Paul. *Alive: The Story of the Andes Survivors*. Philadelphia: J. B. Lippincott, 1974.

Ritterbusch, Dale. "After Shakespeare's 'When Icicles Hang by the Wall.'" *War, Literature & the Arts: An International Journal of the Humanities* 28 (2016).

———. "Ice Bowl." *War, Literature & the Arts: An International Journal of the Humanities* 24 (2012).

———. "The Monuments We Swear By," unpublished.

Rosenthal, Chuck. *Experiments with Life and Deaf*. Venice, CA: Hollyridge Press, April 1, 2007.

Scott, Robert. *The Voyage of the Discovery*, qtd. in Pyne.

Service, Robert W. "The Cremation of Sam McGee." Poetry Foundation. https://www.poetryfoundation.org/poems-and-poets/poems/detail/45081.

Shelley, Mary. *Frankenstein*. New York: Penguin Classics, 1985.

Silber, Joan. "About My Aunt." *Tin House*, 2014.

Solzhenitsyn, Aleksandr. *The Gulag Archipelago, III–IV*. New York: Harper & Row, 1975.

———. *One Day in the Life of Ivan Denisovich*. New York: Bantam, 1963.

Stenta, Gregory. "Lecture on Stalingrad," *War, Literature & the Arts: An International Journal of the Humanities* 24 (2012).

Stewart, George R. *Ordeal by Hunger: The Story of the Donner Party*. New York: Henry Holt, 1936.

Stoutenberg, Adrien. *Some Haystacks Don't Even Have Any Needle*. New York: Lothrop, Lee & Shepard Co., 1969.

Streever, Bill. *Cold: Adventures in the World's Frozen Places*. New York: Little, Brown, 2009.

Sukach, M. K. "Stalled." In *hypothetically speaking*. Farmington, ME: Encircle Publications, 2016.

Thoreau, Henry David. *The Heart of Thoreau's Journals*. New York: Dover, 1961.

Trager, James. *The People's Chronology: A Year-By-Year Record of Human Events from Prehistory to the Present*. New York: Henry Holt, 1994.

Tunstall, Graydon A. *Blood on the Snow: The Carpathian Winter War of 1915*. Lawrence: University of Kansas Press, 2010.